NIST GCR 03-852

Simultaneous Measurements of Drop Size and Velocity in Large-Scale Sprinkler Flows Using Particle Tracking and Laser-Induced Fluorescence

Prepared for
U.S. Department of Commerce
Building and Fire Research Laboratory
National Institute of Standards and Technology
Gaithersburg, MD 20899-8662

I0493826

July 2003

U.S. Department of Commerce
Donald L. Evans, Secretary

Technology Administration
Phillip J. Bond, Under Secretary for Technology

National Institute of Standards and Technology
Arden *L. Bement, Jr., Director*

Acknowledgements

The authors are grateful for the support of this work by the National Institute of Standards and Technology, Building and Fire Research Laboratory, via Grant #60NANB8D0080. The efforts of the NIST Scientific Officer, Daniel Madrzykowski of the Fire Research Division, are greatly appreciated. The authors also extend their thanks to Kazuki Shiozawa, for automating the image analysis and lending assistance during the conduct of experiments.

Simultaneous Measurements of Drop Size and Velocity in Large-Scale Sprinkler Flows Using Particle Tracking and Laser-Induced Fluorescence

Anthony D. Putorti Jr.[1], David Everest, and Arvind Atreya[2]

Department of Mechanical Engineering and Applied Mechanics
University of Michigan, Ann Arbor, MI 48109-2125

Abstract: This paper reports an experimental technique that is developed for instantaneous planar measurement of droplet size and velocity for dilute sprays in general and sprinkler sprays in particular. This particle tracking technique relies on photographic measurements of two-color fluorescence or fluorescence and Mie scattering from droplets to determine their size and velocity. Measurements were made in a plane that passes through the vertical axis of symmetry of an axis-symmetric sprinkler spray. Drop velocities and drop sizes down to 200 μm were measured from the digitized double-exposed photographs of sprinkler sprays. The directional ambiguity was resolved by color differentiation. A combination of fluorescence and Mie scattering was investigated for a 250 mm × 350 mm measurement area, while two-color fluorescence was investigated for 460 mm × 540 mm measurement area. Dye selection and concentrations were determined by testing the fluorescence output of various water tracer dyes. Collection optics and laser powers were varied to optimize the color differentiation and maximize the resolution for drop size measurements.

Keywords: Drop size measurements, PIV, Sprinklers, Laser-Induced Fluorescence

1. Introduction

Water spray sprinklers are the most commonly used automatic fire protection systems in buildings ranging from small offices to large warehouses. For effective fire suppression, the sprinkler water must reach the burning surface. An optimum sprinkler system, for a given application, is one that provides the maximum fraction of water delivered by the sprinkler(s) to the burning surfaces and suppresses the fire in the shortest time after its initiation. The design of such a sprinkler system depends on the geometrical relationship between the sprinkler(s) and the fire source and its heat release rate, the geometry of the room and its ventilation conditions and the sprinkler spray characteristics. Given the complexity of the problem, the optimization, as well as the evaluation, of various sprinkler systems is most cost-effectively accomplished via computer models that can calculate the fire and sprinkler induced flows for different geometries. Such models have been developed at BFRL (NIST) by McGrattan, Baum, Rehm, et. al.[1, 2] For experimental validation and input data for these models, instantaneous field measurements are needed on drop size distribution, drop velocity, sprinkler induced flows, and the actual delivered density. This paper develops a laser-based technique to provide the drop size distribution and drop velocity data.

A number of standard techniques and instruments are available for measuring drop size distributions in sprays such as Phase-Doppler Interferometry (PDI)[3, 4] and the Optical Array Probe (OAP) by Particle Measuring Systems[3, 5, 6] These techniques are suitable for measuring a single point or an array of points, but do not provide the instantaneous spatial drop distribution in the spray. Consequently, a large number of measurements must be taken at different points to determine the drop-size distributions and the spray must be considered time invariant. To overcome this difficulty, an instantaneous large-scale planar measurement technique was designed and measurements were made along the vertical axis of an axis-symmetric sprinkler spray.

Different techniques of simultaneous planar measurement of droplet size and velocity have recently appeared in the literature. Kadambi et al.[7] have identified the errors associated with particle size measurements from Particle

---

[1] Current address: Fire Research Division, National Institute of Standards and Technology, Gaithersburg, MD 20899

[2] Corresponding author.

[3] Certain commercial equipment, instruments, or materials are identified in this paper to adequately specify the experimental procedure. Such identification does not imply recommendation or endorsement by the National Institute of Standards and Technology, nor does it imply that the materials or equipment are necessarily the best available for this purpose.

Image Velocimetry (PIV) images. Herpfer and Jeng [8] have introduced streak PIV for planar measurements of droplet sizes and velocities. Domann and Hardalupas [9, 10] have investigated fluorescing droplet intensity profiles and the effects of dye concentration. Cao et al.[11] and Everest et al[20] have used planar laser-induced fluorescence for measurements of droplet size and Particle Tracking Velocimetry (PTV) for measuring drop velocities. The PTV technique is similar to the PIV technique and is useful when the density of seed particles is low. In this paper, we present a Particle Tracking Velocimetry and Imaging (PTVI) technique, similar to References [11 & 20][11,20], that relies on taking instantaneous double-exposed color photographs of the spray and using them to obtain both velocities and particle sizes. Since the double-exposed photographs are created by laser shots of two different wavelengths, the color differentiation helps resolve the flow direction.

2. Experimental Technique

2.1 Droplet Sizing Considerations

Sizing and tracking a particle in a large field of view (FOV) requires simultaneously satisfying disparate requirements of low magnification and high spatial resolution. For proper sizing, the spatial resolution of the optical, film, and digital components must result in a fully resolvable drop, indicating as large a magnification as possible. Meeting the spatial requirements for drop sizing results in limiting the FOV and a subsequent loss in the dynamic velocity resolution (DVR) and the dynamic spatial resolution (DSR), as defined by Adrian[12]. Greater than optimal magnification required for meeting the sizing requirements reduces the spatial and velocity spectrums over which the velocity measurements can be made. This can be seen by considering the DVR×DSR product. Clearly, a large value of this product is desirable. The DVR×DSR product is proportional to the dimensionless constant L/l that is characteristic of the imaging/recording system. Here, L is a characteristic dimension that defines the format of the recording medium and l is the smallest observable length scale. For PIV applications, 'l' may be as small as two pixels for a digital recording media or 4/(line pairs per mm) for a film recording medium yielding a large DVR×DSR product. However, for simultaneous particle sizing, 'l' is determined by the smallest particle size that needs to be resolved and the required accuracy; it is at least twice the requirements for PIV. Thus, to maintain a large DVR×DSR it is necessary to increase 'L', i.e. a high resolution large format imaging. A medium format film camera was used to satisfy these disparate requirements. Also, particle tracking is used to measure the velocity and color differentiation was used to resolve the directional ambiguity.

The variables that must be considered for particle size measurements by the PTVI technique are schematically shown in Figure 1. First, the thickness of the laser sheets relative to the droplet size and the variation of the laser intensity across this thickness are important. Often the laser sheet thickness for PIV or PTV applications is determined by the transverse component of the particle velocity and the time separation between the two laser pulses. For the worst case scenario of equal in-plane and transverse velocities, the minimum beam sheet thickness (as defined by 50% intensity or full width at half maximum, FWHM) should be at least three times the maximum particle diameter. Thus, for a 3 mm sprinkler droplet, the laser sheet should be at least 10 mm thick[4]. Since the light intensity in Gaussian sheets falls to about 50% of the maximum at the edges, the image of a large droplet at the edge may be smaller than the image of the same size droplet at the center. This error must be quantified for drop size measurements. Next, the effect of the depth of focus (DOF) must be quantified. If the droplet is out of the depth of focus, it appears large because of blurring. Thus, the DOF and the beam sheet thickness should be approximately equal to enable the increase and the decrease in the particle image size to roughly cancel each other out. Increasing the DOF larger than the beam sheet thickness is not useful because it requires larger f#[s] that only serve to reduce the amount of light captured by the camera and increase the error due to diffraction limitation of the optics. It is also possible to use a laser that produces a top hat type energy profile. In fact, in many Q-switched pulsed lasers, the initially gaussian laser beam profile changes to approximately a top hat profile by the time it reaches the measurement region. Since the energy level of the top hat beam is relatively constant across the sheet thickness, the droplets imaged at the edge of the sheet would be expected to

---

[4] The choice of the beam sheet thickness is also important to ensure that sufficient numbers of droplets are present in the measurement volume to obtain a representative sample. The particle seeding density in these experiments is not externally determined leaving beam sheet thickness as the primary control variable. As will be seen later, the criteria used above yields adequate number of vectors to define the velocity field.

be larger due to blurring. All these questions and those regarding the ability to accurately measure drops sizes by recording the scattering signal are addressed below.

Kadambi et al.[7] considered some of the above effects for determining the particle size from the image. In their experiments, light scattering from solid particles suspended in a liquid was captured. They studied the effects of spatial resolution, light sheet intensity distribution, and depth of field (DOF) on particle size measurements. They observed that the particle image must be greater than 3 pixels in diameter to determine its size with reasonable accuracy. For quantifying the light sheet intensity distribution effect and the DOF effect, they used a 200 µm particle, a DOF of 200 µm, and a beam sheet thickness of 1600 µm. By moving the focused particle through the light sheet, they observed a decrease in the measured diameter of approximately 9% and a roughly 50% decrease in the scattered light intensity at the edges of the light sheet relative to the center. In another experiment, they traversed the imaging camera while holding the particle and laser sheet fixed to determine the effect of DOF. It was observed that the particle diameter increased by approximately 9%, while the intensity level again decreased by roughly 50%. Thus, when a particle is not centered in the light sheet and the focal plane, DOF and beam sheet intensity effects approximately cancel each other out resulting in a constant particle diameter. Their results show a standard deviation of 7% and pertain to measurements from the image formed by light scattered by the particles. Their results also indicate that it would be advantageous to have DOF $\leq$ FWHM. (Recall that FWHM $\geq 3 \times d_{DROP}$ for PIV requirements.) While Kadambi et al.[7] and Adrian's[12] results are very insightful, scattering measurements did not yield reliable droplet sizes in the present work. Thus, planar fluorescence imaging along with scattering, or two-color fluorescence imaging, was used to make drop size and velocity measurements in a large FOV with low-density sprinkler spray.

In order to estimate the effects of the experimental setup on the ability to produce suitable droplet images, it would be useful to have expressions for the light signals from the droplets. The fluorescence emission signal 'F' for droplets is given by the equation:

Equation 1

$$F \propto \frac{\phi I_0}{\lambda} \left(\frac{m}{(m+1)f\#} \right)^2 V f(c) f(R)$$

Here, 'F' is proportional to the fluorescence yield 'ϕ', intensity of light incident on the droplet 'I_0', wavelength of emitted light 'λ', and the volume of the illuminated droplet 'V'. The signal is nonlinearly related to the magnification 'm', f#, the dye concentration function 'f (c)' and the film response function 'f (R)'. The scattering signal 'S' from large water drops in a low-density spray is given by:

Equation 2

$$S \propto \frac{I_0}{\lambda} \left(\frac{m}{(m+1)f\#} \right)^2 A f(\tau) f(R)$$

Where the signal 'S' is linearly related to the incident intensity, 'I_0', the wavelength of scattered light, 'λ', and the area of the illuminated drop, 'A'. It is nonlinearly related to the magnification 'm', f#, the angle of incidence of a light ray on the drop 'f (τ)', and the film response function 'f (R)'. For both cases, the signal quality is improved by using higher magnifications and smaller f#s. However, to meet the FOV requirements the magnification is fixed at approximately 0.1, and only the f# can be reduced.

The response function 'f(R)' of the film is dependent on the spectral and exposure response characteristics of the film emulsions. Characteristic sensitivity curves for Kodak Pro film were used to define the film spectral response function. These curves indicate that the film is approximately 30 times more sensitive to 588 nm (rhodamine fluorescence), 434 nm (stilbene fluorescence), and 532 nm (scattering) than for 355 nm UV scattered light. The increased energy in the shorter UV wavelength improves the signal but lower 355nm laser power offsets that gain. As a result, 355nm scattering was observed only in a few of the 250 mm x 350 mm images, and in none of the 460 mm x 540 mm images. Similar results are obtained from Fuji color films.

It is also instructive to qualitatively examine how various signal levels change with the choice of f# of the optics. Two formulas taken from Adrian[12] are useful:

Equation 3

$$\frac{d_{IMAGE}}{d_{DROP}} = m\left(1 + \left(2.44\frac{(1+m)f\#\lambda}{md_{DROP}}\right)^2\right)^{\frac{1}{2}}$$

Equation 4

$$DOF = 4\left(\frac{(1+m)f\#}{m}\right)^2\lambda$$

Equation 3 accounts for the diffraction limitation of the optics, whereas, Equation 4 provides an estimate of the depth of field (DOF). Assuming a magnification 'm' of 0.1, Equation 1 through Equation 4 can be used to study the variation with the f# of the recording optics. These equations are shown plotted in Figure 2. The red curves correspond to 355 nm and the blue curves correspond to 532 nm excitation. The values of $\varphi f(c)f(R)$ and $f(\tau)f(R)$ were assumed such that the relative magnitudes of fluorescence and scattering signals are comparable. The objective is to understand the dependence of the signals on the optics parameters.

For a given droplet size, dye concentration and incident intensity, Equation 1&2 shows how the fluorescence and scattering intensities vary with the f#. Clearly, as the f# is increased, the intensity drops sharply as $1/f\#^2$. Consequently, a low f# is needed to increase the signal levels. The curves corresponding to Equation 3 show the %error due to diffraction limitation of the optics for a 0.5 mm drop. These appear acceptable for drops >0.5 mm diameter, but the error increases sharply to 50% for 0.1 mm droplets imaged using 532 nm light and an f# of 8.0. As expected, the diffraction error for 355 nm images is significantly less than for 532 nm images.

The increase in the DOF with the f# is also shown. DOF must be appropriately chosen based on drop sizes and the beam width. The primary balance that needs to be achieved is between the fluorescence & scattering intensity and the DOF. It appears that the choice of an f# around 3 (say 2.8) is appropriate for these parameters. However, if the maximum droplet size to be measured is much larger, on the order of 3 mm for example, the beam sheet thickness and DOF will need to be larger, making a larger f# necessary. In this case, the error in droplet size measurement due to diffraction effects will increase, but are still less than 10% for 200 µm drop sizes at f#=5.6. Adequate signal for droplet detection and imaging must be verified, however, given the decrease in aperture size. Figure 2 indicates very small fluorescence and scattering intensities at larger f#s, however, it is important to remember that these are relative signal values. As noted above, the values of parameters in Equation 1 & 2, were arbitrarily chosen. Thus, the detectability will be determined experimentally in the sections to follow.

With these considerations, experiments were conducted to identify the effect of the imaging system resolution, fluorescent dye concentration, variations in laser sheet thickness, and laser sheet intensity on drop size measurements. The results of these experiments were used to quantify the ability of the measuring system to accurately characterize the spray velocities and particle size while maximizing FOV.

2.2 Experimental Setup

Two methods are possible to resolve the directional ambiguity in the double-exposed, high-resolution color photographs – first color differentiation by fluorescence and scattering and the second color differentiation by two different fluorescent colors. Thus, two experimental setups were investigated to determine their feasibility. In the *fluorescence/scattering method*, the two images of the droplet on the film consist of one from the fluorescing dye and the other from scattering. In the *dual-fluorescence method*, both images are of the fluorescing dyes, but at two different wavelengths.

2.2.1 Fluorescence/Scattering Method

The flourescence/scattering method was tested in a smaller enclosure with a medium power laser at the University of Michigan. Figure 3 schematically illustrates the experimental layout used for the droplet velocity and size distribution measurements. Water exits through a 8.49 mm ± 0.03 mm diameter (changeable) nozzle at 19.0 L/min ± 0.76 L/min (5 gallons/min ± 0.2 gallons/min) and impinges on a conical strike plate creating the sprinkler spray. A fluorescent tracer dye (rhodamine) is injected into the water far upstream of the nozzle. A dual-pulse Nd:YAG laser provides two laser beams that are formed into light sheets by a single f-6.35 mm cylindrical plano-concave lens. The leading beam sheet, formed by the third harmonic (355 nm) occurs Δt milliseconds (Δt determined by PTV requirements) prior to the lagging beam sheet, which results from the second harmonic (532 nm) of the second ND:YAG laser. The laser energies were approximately 130 mJ/pulse at 355 nm and 230 mJ/pulse at 532 nm and the laser sheet is approximately 5mm thick. Fluorescence and Mie scattered light from the two laser pulses is captured by a 35 mm color film camera using a 50 mm f/2.8 Nikon lens with 800 ASA Fujicolor film. Since the framing rate of the film camera is too low to obtain images on separate frames, the photographs were double exposed. Reddish yellow fluorescent images were used for drop size measurements and time discrimination was obtained by fluorescent and green scattering images created by the two laser pulses firing at two different wavelengths. To control the scattering signal intensity at 532 nm, a Quantaray R2 red filter or a Schott OG530 was used to allow only the fluorescence signal and some portion of the scattered signal to reach the film. Images of water droplets in a region approximately 250 mm × 350 mm and 300 mm downstream of the strike plate were taken at a magnification of 0.1 to characterize the droplet size and velocity. The negatives were digitized with a 2000 dpi × 4000 dpi Polaroid film scanner that resulted in a digital resolution of 71μm in the droplet plane/pixel (based on scattering from two fibers at a known distance apart). The optical spatial resolution of the imaging system was also measured by using a standard USAF 1951 resolution chart. Images were obtained at a magnification of 0.1 for various f#s. The best resolution obtained was for an f# of 2.8 at 3.17 lp/mm which corresponds to resolving a158μm thick line in the droplet plane. A line of this size corresponds to roughly 2 pixels in the image. Assuming that 3×3 pixels are needed to measure the drop size with reasonable accuracy, we obtain the minimum measurable droplet diameter of approximately 200 μm.

The digitized images were processed using the TSI Insight and SCION Image software for determining the droplet velocities and droplet sizes respectively. Fluorescence images were used to determine droplet sizes. These images had sufficient spatial resolution for sizing 200 μm diameter droplets, as well as color-differentiation for resolving the directional ambiguity in velocity measurements. For a laser pulse separation of Δt =1 ms, velocities in the range of 0.2 m/s (from minimum separation between small drops) to 10 m/s (using five times the diameter of large drops[12]) could be measured.

2.2.2 Two-color Fluorescence Method

The Two-color Fluorescence Method was tested at NIST in a larger enclosure with a high power laser. The setup of the two-color method uses similar but different components than the fluorescence/scattering method. Water is discharged through a 4 mm or 6 mm diameter nozzle and hits a 90 degree symmetrical strike plate. The water is recycled via a holding tank and pumps, so the dyes are dissolved in water before the experiments to a concentration of approximately 3.3 mg/L of rhodamine and 10 mg/L of stilbene. Dual Nd:YAG lasers operating at 10 Hz produce co-linear beams, with the leading 10 ns pulse at 532 nm, followed by a 355 nm pulse of 10 ns duration. The laser energies are approximately 400 mJ/pulse at 355 nm and 700 mJ/pulse at 532 nm. A laser sheet of approximately 13 mm in thickness is produced by a cylindrical plano-concave lens with a nominal focal length f=−37.5 which was coated for both wavelengths. The sheet was limited to approximately 10 mm in thickness by installing an 8 mm wide slit after the sheet forming optic.

Images of a 460 mm by 540 mm region downstream of the nozzle were taken at a magnification of 0.13 using a Mamiya RZ series medium format camera (70 mm by 60 mm film) with a 210 mm f/4.5 achromatic lens and Fujicolor NPZ 800 color film. Since the frame rate of the camera is slow compared to the laser frequency, the shutter was left open for one laser beam shot pair thereby double exposing the film. A 532 nm notch filter is installed on the camera to exclude scattering from the 532 nm laser. The rhodamine fluorescence in response to the 532 nm pulse results in yellow droplet images, whereas, the stilbene fluorescence in response to the 355 nm pulse results in blue droplet images. The best results were obtained by using a f# of 5.6.

Images were analyzed by using a custom analysis macro running in ImagePro Plus software. Droplet sizes were determined from the blue fluorescent image, and the droplet velocities determined from the yellow and blue images.

2.3 Determination of Fluorescent Dye Concentrations

Proper choice of the fluorescent dye and its concentration must be made to obtain the best signal. The optimum droplet dye concentration was determined by placing dye/water solutions of various concentrations in a Perkins-Elmer luminescence spectrometer (fluorimeter.) The device illuminates the solution sample with a user selectable narrowband light source, and scans the fluorescent output over a large range of wavelengths, recording the intensity at each wavelength. The sample is held within a square cross section cuvet, constructed of polymethyl-methacrylate, which is nearly transparent and consistently transmissive at the wavelengths of interest. The transmission path length is nominally 10 mm.

Solutions of rhodamine, stilbene, and fluorescein were tested in the fluorimeter. The Rhodamine WT dye was provided in a 2.5% by mass in water concentrate, manufactured by Kingscote chemical company. The Stilbene 420 dye was manufactured by Exciton, and designated as non-laser grade, 99% pure stilbene. The fluorescein (uranine) dye was provided in a 7.5% by mass water concentrate, also manufactured by Kingscote chemical company. The results are shown in figure 4 where relative intensity, integrated over all emission wavelengths, is plotted against concentration. From the graph, it can be seen that the best concentrations to use are approximately 3 mg/L of rhodamine in water, 10 mg/L of stilbene in water, and 100 mg/L of fluorescein in water. Note that the maximum response of the stilbene dye is nearly an order of magnitude greater than the response of the rhodamine dye, and that fluorescein reaches its peak at an order of magnitude greater concentration than rhodamine and stilbene. Chemical and identification data for each of the dyes is provided in Table 1. Unless stated otherwise, the rhodamine used in this work is Rhodamine WT.

Table 1. Fluorescent dye information.

Dye	CAS#	Molecular Formula	Molecular Weight	Notes
Rhodamine 6G	989-3838	$C_{28} H_{30} N_2 O_3 HCl$	478.4	
Rhodamine WT	37299-86-8	$C_{29} H_{29} N_2 O_5 Cl.2 Na$	579	Liquid Concentrate 2.5% by Mass
Fluorescein (Uranine)	518-47-8	$C_{20} H_{12} O_5$	332.15	Liquid Concentrate 7.5% by Mass
Stilbene 420	27344-41-8	$C_{28} H_{20} O_6 S_2 Na_2$	562.56	99% pure powder

The fluorimeter is especially useful since it gives the spectral response of the dye to a given excitation wavelength. Spectral response curves for the dyes, in response to 355 nm and 532 nm excitation signals, are shown in figure 5 through figure 9. The maximum emission wavelength varies somewhat with concentration. Two of the dyes, rhadamine and fluorescein respond to both excitation wavelengths, while stilbene does not fluoresce when exposed to 532 nm. These plots, along with figure 4, suggest an optimal combination of dyes for two-color droplet differentiation. If rhodamine and stilbene are used together with excitation wavelengths of 355 nm and 532 nm, the 355 nm light will cause a very strong emission from the stilbene in blue, and a much weaker emission in yellow. With a 532 nm excitation signal, a strong emission from rhodamine will occur in yellow, with no response from stilbene. The final result from the rhodamine-stilbene-water mixture would be a blue emission from 355 nm, and a yellow emission from 532 nm excitation.

Signal intensities from solutions containing both rhodamine and stilbene are shown in figure 10. Note that the emissions from the combination solutions are less than the emissions from the solution with stilbene alone (superposition is not valid), even though rhodamine fluoresces in response to 355 nm light. From the data in figure 11, it can be seen that the primary emission wavelength from the stilbene dye is absorbed by the rhodamine dye. Therefore, when the two dyes are in solution together, the rhodamine will attenuate some portion of the signal from the stilbene. Stilbene, however, has no absorption effect on the fluorescence signal from rhodamine, nor does it fluoresce in response to 532 nm light. For these reasons, the stilbene concentration will be set to its emission maximum (10 mg/L), while the rhodamine concentration will be set to the lower concentration side (3 mg/L) of the emission maximum. This should provide the maximum possible signal.

The spectral responses for the dye solutions can also be corrected for the effect of the 532 nm notch filter (Figure 12), since it will remove some part of the fluorescent signal. The responses of the dye solutions with the calculated response of the notch filter are shown in figures 13 and figure 14. The filter has little effect on the integrated intensity function for 532 nm excitation since the portion of fluorescent emission spectrum near 532 nm is small. The filter does have an effect on the emissions from 355 nm excitation, however, since the fluorescent emission spectrum overlaps the region about 532 nm where the filter is active. Since the fluorescence signal from stilbene is very strong, however, there is still sufficient signal for imaging.

In a different set of experiments, various concentrations of rhodamine and fluorescein dyes were qualitatively tested by passing a beam through a glass tube filled with the mixture. The fluorescence emissions were imaged onto videotape by a Cohu CCD camera with a f/2.8 lens and Schott OG530 filter, and analyzed using a Data Translation frame-grabber board and ImagePro software. The OG530 filter was chosen to allow some 532 nm scattering to be imaged. The results were consistent with those obtained from the spectrometer data shown in Figure 4. These experiments confirm that small-scale spectrometer data is applicable to the larger-scale experiments. 532 nm scattering from air bubbles was also investigated, and observed to have a signal that is equal to 75% of the net fluorescence signal at a concentration of 6.6 mg/L. Since the scattering of a water drop in air is expected to be an order of magnitude greater than an air bubble in water, clearly scattering will be dominant in the sprinkler flow when fluoroscein dye is used. When rhodamine dye is used, test results indicate that scattering and fluorescence will be of the same order of magnitude with the OG530 filter. Absorption was also qualitatively measured and observations indicated that the rhodamine dye is more effective than fluorescein for absorbing both excitation wavelengths. The 532nm excitation was more effectively absorbed than the 355nm excitation. Beyond 100 mg/L, rhodamine and fluorescein have a high absorption that reduces the fluorescence emission. It is fortunate that the fluorescence maximum occurs prior to the absorbance maximum, because it enables illuminating the entire droplet field containing the optimum dye concentration.

For imaging droplets using the fluorescence/scattering technique, rhodamine dye was used at a concentration of 3.3mg/L. If used at higher concentrations, Figure 4 indicates that the signal would not increase and due to absorption of the laser beam, the entire droplet field may not be uniformly illuminated. If used at lower concentrations, the UV fluorescence signal becomes too weak. As shown below, scattering appears to dominate the drop signal except when it is fully in the laser sheet, where the scattering vs. fluorescence signal depends on the scattering angle.

2.4 Image Resolution

Assuming that the signal from the droplets is sufficient to form detectable images, the uncertainty of size measurements may be estimated by examining the resolution of the system. Resolution may be measured with the USAF 1951 resolution chart, which consists of an array of parallel alternating light and dark bars of various spatial frequencies. (figure 15) A line pair per mm of resolution is defined as one dark bar and one light bar, and abbreviated lp/mm. The chart is photographed and digitized with the imaging system, and the magnified images examined. The images can be viewed with the eye, and the set of differentiable lines with the highest spatial frequency is said to be the resolution limit of the system. Since the final component (retina) of this measurement is quite variable from person to person, the method is subjective. In order to make the method objective, the contrast between the light and dark bars is measured as the spatial frequency of the lines increases. Contrast is defined as[13]:

Equation 5

$$C(v) = \frac{I_{max} - I_{min}}{I_{max} + I_{min}}$$

where: I_{max}= intensity level at brightest location of light bar
I_{min}= intensity level at darkest location of dark bar

ν = spatial frequency (lp/mm)

As the optical system measures lines of gradually increasing frequency, the system will reach a point where it can no longer fully capture the transitions from light to dark and dark to light. The contrast between light and dark decreases, until $C(\nu)$ in Equation 5 reaches zero at the cutoff frequency. For diffraction limited (perfect) systems, the cutoff frequency[13], RES_C, expressed in lp/mm, is given in Equation 6.

Equation 6

$$RES_C = [\lambda \ (f\#)]^{-1}$$

The function describing the performance of the system is the modulation transfer function (MTF). The MTF illustrates the reduction in contrast of the image as the spatial frequency increases, and is defined as:

Equation 7

$$MTF(\nu) = \frac{C(\nu)}{C(0)}$$

The MTF can be used to quantify the performance of each component in the imaging system, and the overall system MTF later determined by calculating the product of the individual MTF functions. In order for the calculation to be valid, the optical components cannot be coherently linked. Since we effectively have a diffuser between each of the groups of lenses (ie. film between the camera lens and scanner lens) our components are not coherently linked. For systems that are coherently linked, a more general optical transfer function with real and imaginary parts is used to account for the phase and magnitude of the light[14].

For a perfect system, ie. diffraction limited with no aberrations, the maximum useful resolution is given by the Rayleigh limit. At the resolution corresponding to the Rayleigh limit, the separation of two images is such that the maximum of one diffraction pattern coincides with the first dark ring of the second diffraction pattern. At this distance apart, the droplet images are degraded, but the presence of two maxima can be clearly seen. The resolution at the Rayleigh limit, RES_R, in lp/mm is[15, 13]:

Equation 8

$$RES_R = [1.22\lambda \ (f\#)]^{-1}$$

For an $f\#$ of 5.6, the resolution of 355 nm excited droplet images (439 nm emission) is 333 lp/mm and the resolution of 532 nm excited droplet images (592 nm emission) is 247 lp/mm. Recall that this is the limitation of a perfect system, which is only limited by the diffraction of light due to the camera aperture. Resolution stated in this manner is most useful when closely spaced objects are observed. Since the droplets to be measured in our system are in a dilute two phase flow, we expect our droplets to be many diameters apart. This representation of resolution is useful for determining the limitations of the measurement system, however, since it also indicates how quickly the system can transition from light to dark areas at the edge of droplets. The value of the MTF is approximately equal to 0.09 at the Rayleigh resolution limit[15].

The system uncertainty is examined using the MTF concept since film and lenses do not have pixel equivalents. The scale of interest in this case is 200 μm, which is the smallest droplet of interest in the current experiments. Thus, a cycle would occur over 400 μm, which is equivalent to 2.5 lp/mm in our measurement region. Assuming a magnification M=0.13, the cycle would be 19 lp/mm on the film. Next we determine the fractional value of the response for each optical component at a spatial frequency of 19 lp/mm from the MTF curve for the component. From Fuji NPZ800 film datasheets[16], the MTF (at 19 lp/mm) = 0.85. For a f/4.0 maximum aperture Mamiya lens, the overall MTF (at 19 lp/mm) is given as 0.71 by Photodo[17]. We have measured the entire system MTF (at 19 lp/mm) = 0.20 for the combination of lens, film, and scanner. Since the MTF curve for the scanner is not available, back calculation yields the MTF (at 19 lp/mm) for the scanner alone as 0.30. It is clear from these calculations that the limiting component in the system is the scanner which has a performance level significantly lower than that of the camera lens and film. In addition, the spatial frequency of interest is less than the Rayleigh criterion, so the system is limited by aberrations or defocusing effects in the system not by diffraction. The MTF calculation is valuable for determining the potential difference in performance of various films. One such comparison showed that there was no resolution penalty for using Fuji 800 ASA film versus Fuji 400 ASA film, but the speed difference resulted in a significant gain in light sensitivity.

While the MTF analysis has indicated that the limiting component in the system is the scanner, the results are not directly applicable to determining how the system resolution limitations affect the uncertainty in droplet sizes. Disagreements exist in the literature as to how the MTF relates to the quality of a complex image.[18] Since the scanner is the limiting factor in the measurements, this implies that the effect of the pixels (spatial sampling frequency) is important. The ability of the finite pixels or pixel equivalents to represent the droplets, especially the edges, is of primary concern. This behavior will have a direct effect on how the edge is represented, and therefore the droplet size. The MTF concept tells us how the system spatially transitions from light to dark and vice versa. The effect of the scanner, due mostly to pixelization, greatly affects our measurements since the pixels are relatively large compared to the smallest droplet size (200 μm), and relative to the film, lens, and aperture capabilities.

The scanning of the resolution chart is examined at the pixel level. The photos are digitized with a 4000 dpi optical scanner, and if we neglect the circle of confusion formed by diffraction limitations of the lens or aperture, or by the image being slightly out of focus, the most detail we can expect from the image is on the order of the pixel size. As the line width begins to approach the pixel size of the scanner, several observations can be made. Clearly, lines can not be resolved if the pixel size is greater than the line width. The maximum performance of the scanner would take place when the line width is the same as the pixel width. In practice, however, the lines are not always aligned with the pixels, the result being a reduction in contrast due to pixel averaging of the line edge over two pixels. Therefore, it is necessary to have each line pair covered by more than 2 pixels, with previous efforts suggesting that 3 to 4 pixels per line pair are necessary. The effects of pixels that overlap line boundaries are shown in figure 16. It can be seen that the signal level drops due to the pixel averaging over the line boundary. This may be partially responsible for the asymmetrical intensity peak on one side of the droplet image, although off-center peaks due to the spatial distribution of fluorescent intensity are usually caused by the internal reflections within the droplet[9]. From figure 16, it appears that the "correct" line width is found by using one pixel in from the edge where the signal rises above the background. In this case, the uncertainty in the line width would be ±2 pixels. By using this calculation method, it should be possible for the film/scanner combination to resolve between 5.1 lp/mm and 6.8 lp/mm in the measurement plane for M=0.13. Measurements from the imaged USAF 1951 resolution chart result in the following approximate values of resolution at the Rayleigh criterion (MTF(0.09)): for 8 bit per color channel scanning the resolution is 3.6 lp/mm at f#=5.6, and 5.0 lp/mm at f#=8.0. For 16 bit per channel scanning, the resolution is 4.0 lp/mm at f#=5.6, and 5.0 lp/mm at f#=8.0. The data is summarized in Table 2. Contrary to the diffraction errors predicted in Equation 3, the measured resolution is higher for the smaller aperture, due to less defocusing error. As can be seen from the f#=5.6 case, scanning the images at higher color bit depths may increase the resolution.

Table 2. Aperture effect on resolution.

f# (camera aperture)	Bits per channel (RGB)	Resolution, film plane (lp/mm)	Resolution, measurement plane, M=0.13, (lp/mm)
5.6	8	27.7	3.6
5.6	16	30.8	4.0
8.0	8	38.5	5.0
8.0	16	38.5	5.0

2.5 Identification of Minimum Signal Levels

It is important to understand how the fluorescence signal intensity varies with the drop size and the f# of the optics. The theoretical curve corresponding to Equation 1 was plotted in Figure 2. Experimentally this was investigated by imaging fluorescence from a 2.8 mm ± 0.14 mm diameter drop. The droplets were centered in both the nominally 5 mm thick laser light sheet and the depth of field of the camera. A long-pass 550 nm filter was used to collect 588 nm rhodamine fluorescence resulting from 355 nm excitation. Images were taken at a magnification 'm' of 0.09 for f#s of

1.2, 2.8, 4.0 and 5.6. The laser intensity was approximately 2.2 times higher than that used in the 250 mm × 350 mm imaging area spray experiments to be described later. The measured change in the relative maximum droplet image intensity (arbitrary units) with f# is shown in Figure 17. The average background signal of 34 was subtracted from the data. The droplet diameters measured from these photographs were 2.7 mm ± 0.27 mm.

The fluorescence intensity was also measured from 201 μm ± 1.34 μm ethanol droplets doped with stilbene and rhodamine dyes using the dual-fluorescence setup. The droplets were illuminated with a 355 nm laser sheet 13.5 mm ± 0.5 mm thick, and centered within the laser sheet and camera depth of focus. A 532 nm notch filter was installed, which was used to remove the 532 nm scattered light. Images were taken at a magnification of approximately 0.20 at f#s of 4.5, 5.6, 6.7, 8.0, and 9.5. The laser intensity was the same as that used in the 460 mm × 540 mm spray experiments to be described later. The image intensity, normalized by the scanner saturation intensity, is shown versus f# in figure 18.

Contrary to our expectation, the measured signals from both size droplets are not linear with $1/f\#^2$ for the entire range. The intensity function appears to be roughly linear at the higher intensity levels (large apertures), but decreases at an accelerating rate as the aperture size decreases. For the 2.8 mm droplets, this behavior can be partly explained by the decreasing depth of field as the aperture increases in size according to Equation 4. For the given magnification and wavelength, the depth of field is greater than ~3 mm for f#s greater than 2.8 (see figure 2). However for f# of 1.2, the depth of field is only 0.5mm, much smaller than the drop diameter. As discussed below, the 355 nm beam sheet FWHM was approximately 3.5 mm for the 2.8 mm droplets, much larger than the depth of field and slightly larger than the drop. The drop is therefore fully illuminated, but the fluorescence outside the depth of field is not in focus and therefore does not increase the signal as anticipated. Hence, the depth of field serves to limit the fluorescence volume used in Equation 1. For drops that are larger than the depth of field, the fluorescence volume is likely proportional to $DOF \times (d_{drop})^2$ rather than $(d_{drop})^3$. Since DOF is also proportional to $f\#^2$, Equation 1 and Equation 4 indicate that for drops larger than DOF, the intensity should not depend on the f#. There is some increase in fluorescence signal at f#s smaller than 2.8, however, as indicated by the data in figure 17, the droplet portion out of the depth of field still makes some contribution to the measured signal.

Examination of the signal levels from the 200 μm nominal diameter droplets provides more information on the subject. Since the droplets are well within the laser sheet and depth of focus for all aperture settings, blurring is not a factor. Instead, it appears that the non-linearity of the intensity function at low signal levels is due to the film response function. A sample film response curve is shown in figure 19. It can be seen that the film response to light is linear over a middle range of exposures (exposure intensity and exposure time are both important), but the slopes change dramatically at low and high exposures, a phenomenon referred to as the reciprocity effect. Therefore, at low signal levels, an increase in signal level has less of an effect on film density than at higher exposure levels. Since we are conducting our experiments in the mid-range of intensities, we will be working within the roughly linear region of the film response function.

In addition to the resolution limited droplet size, the minimum detectable drop size must also be determined or estimated. It can be estimated by calculating the minimum measurable fluorescence signal required to illuminate a pixel in the digitized photograph. This signal can be generated by a small diameter drop or by the edges of a larger droplet. A relationship, based on Equation 1, may be used to estimate the minimum detectable drop size. Since many of the factors in Equation 1 have not been measured individually, the expression can be rearranged to combine them into a constant of proportionality, γ, assuming they are invariant as follows:

Equation 9

$$F = \gamma \left(\frac{I_0 V}{\lambda} \right) \left(\frac{m}{(m+1) f\#} \right)^2$$

If the information on functions for f (c) and f (R) is available, they may be separated from 'γ' in order to study the effect of changing these variables on the fluorescence signal. This may be especially useful for studying the performance of a measurement system at low fluorescent intensities where the film response function becomes non-linear.

If we assume that a signal to noise ratio (SNR) ≥ 2 is measurable, then we need to quantify the noise. From the 2.8 mm drop imaged in the fluorescence/scattering setup, it was determined that the noise in the background signal is about 10% of the average background level of 34. This indicates that intensity levels less than 7 units above the background will have a signal to noise ratio (SNR) less than 2 and by definition not measurable. For small drops, the fluorescence volume in Equation 1 is proportional to the pixel area×drop diameter until the DOF limit is reached. Data from the curve fit in Figure 17 represents Equation 1 and can be used to calculate the value of γ, and Equation 9 used to find the expected signal for a small drop. Calculations show that for f# = 2.8, droplets smaller than 104µm will give a fluorescence signal below the desired SNR of two. Similarly for f# = 4, 213µm defines the lower limit of detectability. In both cases, the limiting fluorescence diameter is greater than the pixel dimension. Equation 9 can be used to estimate the lower identifiable limits in the spray experiments where the beam sheet intensity was 2.2 times less and the magnification was 0.1. The lower limit under those conditions is 189µm for f# of 2.8, and 387µm for f# of 4. The smallest detectable drop size will be different for actual sprinkler measurements versus the above calculation or for streams of calibration droplets for the following reasons. Since the sprinkler will contain many droplets that are not in the sheet, these out of plane droplets will effectively raise the noise level and the intensity of the background. When the background intensity is defined to exclude the out-of-plane droplets, very small in plane droplets that may have been detectable will have the same intensity as the noise and will be excluded. This subject will be discussed in depth in the droplet sizing method section.

The f# experiment was repeated for scattering of the 532nm beam. The experimental parameters were the same as above except both fluorescence and scattering from the drop contributed to the signal. The drop size estimated from the images was 3.3 mm ± 13%, larger than the expected 2.8mm diameter. Scattering was observed from the portion of the drop on which the laser beam was incident. The size of this region changed with f#, but the maximum scattering intensity was insensitive to the increase in f#, due to saturation of the film. Thus, it was concluded that the scattering from the drops at 532nm excitation is very useful in identifying the location of drops, but not so in identifying the size.

To determine if the 355nm excitation and 532nm excitation would result in the same drop size, a series of images were taken with long pass filters used to cutoff the short wavelengths. With no filter primarily scattering from 532nm is observed. The 550nm filter adequately blocks the scattering and shows a round drop (eccentricity = 1.04), 3.0mm ±10% in diameter for 532nm excitation and a slightly smaller drop (eccentricity = 1.05), 2.7mm ±10% in diameter for 355nm excitation. The maximum fluorescent signal for 532nm excitation was about 16% larger than that for the corresponding 355nm excitation.

The minimum detectable droplet size was also estimated for the two-color fluorescence setup. From measurements of 200 µm water/dye droplets in a 355 nm laser sheet and M=0.13, the value of γ=4.95x10^{12} m^{-3} for an f-stop of 8.0, and γ=4.00x10^{12} m^{-3} for an f-stop of 5.6. This data is also plotted in figure 18 along with the alcohol droplet data. Note that 'F' has been normalized by the maximum intensity of a film pixel. According to Equation 9, the two values of γ should be equal, however experimental uncertainty and some non-linearity in the 1/f#^{2} relationship is present. Note that it is inappropriate to compare the magnitude of the intensities of the water versus alcohol due to differences in the experimental setup and dye concentration. For all factors being equal, the dyes in alcohol were found to produce a greater fluorescent intensity than in water.

The minimum detectable drop size for the two-color fluorescent setup is based on a more severe criteria than an SNR≥2. In this case, the droplet intensity must be greater than 2 times the background level. From Equation 9, the minimum detectable drop size for the two-color fluorescent setup would be approximately 130 µm. (Normalized background noise level was 18/256, so the drop must have a normalized intensity of 36/256 for detection) Based on the background cutoff used in sprinkler sprays, however, the minimum detectable droplet size would be approximately 132 µm. (Cutoff level of 39/256 was defined in the droplet sizing and velocity algorithm to be discussed later.) This estimate is based on dye concentrations, magnification, laser powers, and film/scanner characteristics that are the same as will be used to measure the sprinkler flows.

There is significant uncertainty in the above estimates since all of the factors in Table 3 are important, and not all of the factors can be well quantified. Therefore, the calibration is important in identifying whether or not the 200 µm droplets are detectable throughout the laser sheet. It would also be useful to have an estimate of the smallest drop detectable, in order to estimate the number of drops present in the sprinkler spray that are smaller than our 200 µm lower bound. These sub-200 µm drops can be estimated if they are detectable, and on the order of the pixel size to reduce the loss from pixel averaging errors.

Table 3. Measurement system calibration parameters.

Laser Power	Film sensitivity	Film resolving power
Magnification	Aperture / sheet thickness	Lens resolution
Droplet size range	Scanner resolving power	Dye concentration

Assuming that the intensity of fluorescence from a 200 µm droplet is adequate for detection, the measurement area and pixel size can be determined. Using the guideline of having the drop covered by a 3 pixel x 3 pixel matrix (M=0.10) should result in an error of approximately 30%, while a 4 pixel by 4 pixel matrix (M=0.13) should result in an error of approximately 25%.

Now, from the relationships for estimating the minimum visible droplet size, as well as the laser profile and depth of field information, an appropriate f-stop can be chosen. This choice will be a compromise since the optimum estimated f-stop from equalizing the depth of field and beam width results in too little light reaching the camera film. Therefore, f# of 1.2 was chosen for the fluorescence/scattering measurements (250 mm x 350 mm), and f# of 5.6 was chosen for the two-color fluorescence (460 mm x 540 mm) measurements. The use of the f-stop of 8.0 was also investigated for the two-color fluorescence measurements, but 5.6 is a more severe case from the focus perspective, and also provides for decreases in laser power over time.

2.6 Effect of Beam Sheet Thickness and Dye Fluorescence onDroplet Size and Velocity Measurements

2.6.1 Fluorescence and Scattering method

The laser beam incident intensity, the illuminated drop volume, and the incidence angle of the ray on the drop are important for determining the intensity of the signal collected from the drop. The incident intensity is a function of the location of the drop in the light sheet, both vertically and in the depth of the sheet. The beam sheet thickness for 355 nm excitation and for various powers of the 532 nm excitation are shown in Figure 20, normalized by the full width at half maximum (FWHM) of 4.4 mm for the 532 nm beam at 220 mJ/pulse. For high laser power, the visible excitation beam sheet is thicker than the 355nm sheet. Thus, 532 nm scattering is seen more often than fluorescence due to 355 nm.

The photographs taken at higher laser power (220 mJ/pulse for 532 nm beam sheet and 130 mJ/pulse for the 355 nm beam sheet) are shown in Figure 21. Here the pulse separation between the two beams is 4 ms. A 2.8 mm drop is traversed in 1 mm increments from behind through the light sheets. These drops were ejected from a 22-gauge hypodermic needle at a rate of 6 drops/s and images were taken with a resolution of 71mm/pixel at a downstream location of about 160 mm. The highest velocity of the drops, accelerated by gravity, is about 1.8 m/s. Fluorescence at 588 nm from 355 nm excitation is observed in the upper drop in each frame, while scattering and fluorescence from 532 nm beam is observed in the lower drop. Filters were not used for taking these images. The scattering of the 532 nm laser beam by the forward edge of the drop is the first and last of the signals observed in the sequence. The largest source of this light is reflected light although signals from the rear and top of the drop indicate the effect of internal reflection. In the sequence of photographs from A to U, the fluorescence from 532 nm appears before the fluorescence from 355 nm due to both a wider beam and the increased response of the dye to 532 nm excitation. The 532 nm drop measurement indicated elongated drops with an average eccentricity of 1.14, while the 355 nm measurements were more round with an eccentricity of 1.07. There is however one problem, color differentiation is not prominent. Both sets of drops appear yellow & the primary distinguishing feature is the green scattering. Thus, while it is possible to use this technique in a spray to determine both size and velocity, color discrimination is expected to be tedious. Consequently, a different dye was sought to fluoresce in a color different than yellow. This led to the use of stilbene which fluoresces in blue in response to 355nm excitation. Also, since scattering often overexposes the film it was decided to eliminate scattering by a notch filter and use two-color fluorescence. This method will be described later.

Another important consideration in determining the beam sheet thickness is the size of the particle that is to be measured. If the sheet is very thin relative to the particle, the probability of measuring the true diameter is small. The measured mean drop diameter is related to the beam sheet thickness by the following formula:

Equation 10

$$D_m = \frac{\left(\left(\frac{2}{3}\right)^{\frac{1}{2}} D_a^2 + D_a b\right)}{D_a + b}$$

Where 'D_m' is the measured mean drop diameter, 'D_a' is the actual drop diameter and 'b' is the beam sheet thickness. The relation is derived from the geometric probability of some portion of the laser sheet illuminating the true diameter of the droplet. If the beam sheet is infinitesimally thin, the measured diameter is 81.6% of the actual diameter. This means that if all the drops were uniformly the same diameter, the images would show a variety of sizes with a mean diameter that is 81.6% of the actual diameter. If the beam sheet is equal to the drop diameter, the measured drop diameter will be 90.8% of the actual diameter, whereas, if b is 10 times the diameter of the drop, the measured average drop diameter will be 98.3% of the true diameter for mono-dispersed drops. When a variety of drop sizes are to be measured instantaneously, it is important that the measured diameter be as nearly equal to the true diameter as possible. This would indicate that a wider beam sheet is better than a thin one. In the fluorescence/scattering spray experiments, the 532 nm beam sheet thickness was maintained at the FWHM of 4.4 mm.

Another reason for a thicker beam sheet is related to the size of the sampling region. Given a number of drops randomly distributed in a volume such that the mean distance between drop denoted is 's', then there is a lower limit to the smallest volume that can be sampled and still return the correct number density, which is $1/s^3$. Sampling with a beam sheet dimension less than the mean distance between drops will result in underestimating the total number of drops and over-sampling the large drops. In the sprinkler experiments, the laser beam was expanded in only one direction, leaving the sheet thickness equal to the initial beam diameter of 5 mm. The water flow rate was calculated from the measured drop diameters and the mean axial velocity of the drops. The calculated water flow rate matched the rotameter flow rate to within the measurement error.

2.6.2 Two-Color Fluorescence

As stated earlier, to improve the color differentiation, a two-color fluorescence method was devised and the effect of droplet beam sheet position was further investigated in a larger (460 mm x 540 mm) area with a medium format camera. This was done to explore the limits of the technique and develop a method for measurements in real sprinklers. The measurement system was assembled in the same configuration as used for the spray experiments. The laser power profiles were determined by projecting the laser sheet on a sheet of paper. The camera on the opposite side imaged the paper, and the scanned image analyzed. The exposure of the film (f#) was set so that the most intense portion of the sheet on the image did not saturate the film. Beam sheet profiles are shown in figure 22 through figure 25 (Y). One set of beam sheet measurements was conducted using the original beam diameters from the lasers. In order to equalize and reduce the beam sheet widths, an 8 mm slit was installed after the sheet forming optic, resulting in the second set of power profiles. The laser power fluxes at the vertical location of zero (where the unexpanded beam would hit) are shown in Table 4. Based on the second set of sheet profiles (with slit installed) and depth of field calculation, the best aperture setting for a magnification of 0.13 would be f# of 9.5. A plot showing the predicted depth of field for various f#s is shown in figure 26.

Table 4. Incident radiation flux at vertical center of laser sheet.

Wavelength (nm)	8 mm slit installed?	Nominal Beam Thickness (mm)	Incident Flux (W/m^2)
355	No	14	315
532	No	11	450
355	Yes	11	250
532	Yes	10	350

In order to provide measurements with low levels of error and documented levels of uncertainty, the measurement system is calibrated with droplet diameters that span the design range of 200 µm to 3000 µm. This size range was chosen based on previous research[6] that indicated that approximately 98% of the water from typical fire sprinklers is contained in droplets larger than 200 µm in diameter. The droplet production device must be capable of delivering the monodisperse droplets through a known and consistent trajectory. Only in this way will it be possible to compare the size of the droplet image to the actual size of the droplet, and gauge the effect of particle location in the laser sheet on the measured droplet diameter. The 3000 µm water drops were produced by a 22 gauge hypodermic needle, carefully cut and honed to produce an opening perpendicular to the shaft of the needle. The needle is fed with an elevated reservoir, using the elevation head to cause flow out of the needle. The syringe is attached to a dial caliper traversing device, constructed to allow the needle to be moved through the thickness of the sheet in 1 mm increments. The apparatus allows the monodisperse droplets to fall with a given vertical trajectory, and the trajectory to be moved in a precise manner across the thickness of the laser sheet. The readability of the caliper is 0.001 in., and the uncertainty in the measurement is $\pm 2.54 \times 10^{-2}$ mm (0.001 in). The droplet size is determined by measuring the mass of the droplets. Rhodamine and stilbene dyes are added to the water prior to filling the elevated reservoir.

A monodisperse stream of 200 μm droplets with consistent trajectory is formed using a vibrating orifice aerosol generator (VOAG) offered by TSI, inc. This device operates by passing water through a vibrating piezoelectric section and out an orifice. The water exits the orifice as a stream, and breaks up into a monodisperse series of droplets, the size of which is governed by the vibration frequency, orifice size, and fluid flow rate. The device will produce droplets on the order of 20 μm to 400 μm in diameter depending on the operation parameters. For water, the maximum droplet size is approximately 200 μm in diameter. In order to produce the 200 μm droplets, the VOAG was operated with a 100 μm diameter orifice, a fluid flow rate of 2.2 cm^3/min, and driven by a sinusoidal square wave vibration frequency of 8.621 kHz with a magnitude of 4V trough to peak.

The VOAG was mounted on a traversing mechanism constructed to allow the droplet stream to traverse the sheet. The traversing mechanism provides 1 mm of movement per revolution, and a linear mm scale for reference. The uncertainty in the measurement is estimated at ± 0.13 mm (0.005 in.) The device was found to be very repeatable during the experiments, with minimal lash in the mechanism. The results of the drop size calibration/validation are stated and discussed later in the drop sizing method section.

2.7 Aperture and Scanner Effects

The final effect of the aperture size on qualitative image quality is shown in figure 27. This figure shows 2410 μm ± 20 μm droplets illuminated by a 355 nm laser sheet. The images were carefully scanned in order to provide the best possible focus from the scanner. It can be seen that the best image is provided at f=8.0, although higher image intensity and only slightly worse droplet boundaries are present at f=5.6 and f=4.0. The droplet clearly has more distortion at f=2.8. The affect of improper focusing by the scanner introduces much more distortion than aperture size, as shown in figure 28. In this figure, two successive scans of the same image are compared. The scanner, which utilizes an auto focus mechanism, appears to have difficulty with photos having large dark areas. Images of actual sprinklers, with large brightly-lit areas, are less susceptible to focus errors. The scanner also has a multiple exposure negative holder, with one position that works much better than the others. The manufacturer suggested disabling the auto focus mechanism, but this did not improve the image quality as determined from the resolution charts and droplet images. Figure 28 also shows how the color differentiation is accomplished by the two-color fluorescence method.

2.8 Droplet Intensity Profile

The intensity profiles across droplet images were measured with the two-color fluorescence setup in order to formulate a method for determining the droplet size. Other researchers have recently predicted and measured two-dimensional fluorescent intensities for spherical droplets, and suggested that droplet intensities are better predicted when the concentrations of fluorescent dye are within a certain range.[9, 10] This range is expected to vary with droplet diameter since droplet fluorescent intensity is proportonal to D^3 when incident light rays are reflected within the droplet a number of times before being absorbed. Thus, if the dye concentration is too high, the incident light rays will be absorbed before they are allowed to bounce around the inside of the drop and induce fluorescence in all areas of the droplet volume, and the fluorescent intensity is predicted to approach proportionality with D^2.

The droplet intensity measured along the horizontal centerline of the droplet image parallel with the incident light rays is shown in figure 29 for a 2921 μm ± 27 μm (2σ statistical) droplet illuminated with 532 nm and 355 nm light. The droplets were located in the center of the beam sheet, and therefore were uniformly illuminated. The droplets were composed of distilled water doped with 3.3 mg/L (5.7×10^{-6} M) of Rhodamine WT and 3.0 mg/L (5.4×10^{-6} M) of stilbene. The plots indicate slightly different droplet diameters as measured by the width at half maximum, a phenomenon at least partially caused by the inability to match the laser profile and camera depth of field for both wavelengths simultaneously.

Normalized emission profiles for the 3 mm nominal diameter drops doped with Rhodamine WT and stilbene are compared to those predicted by Domann and Hardalupas[9] for 200 μm drops doped with Rhodamine 6G in figure 30 and figure 31. Note that the fluorescent emission profile from Rhodamine WT is a top hat and doesn't match the predicted profiles. The measured stilbene emission profile, however, is qualitatively similar to the predicted high concentration profile for Rhodamine 6G, but with a low concentration of stilbene. In this case, the 500 mg/L Rhodamine 6G prediction is in reasonable agreement with the 3.0 mg/L stilbene data. This is unexpected since there is a three order of magnitude difference in the molar concentrations, but only one order of magnitude difference in diameter between the two. The difference in droplet diameter will have some effect, but does not account for the agreement given the predicted[10] effect of increasing droplet diameter.

The 3 mm nominal diameter droplet images previously discussed do not saturated the photographic film. However, in application of the measurement method, saturation of the film with the signals from large droplets is necessary in order to have sufficient intensity above the background and noise to image the 200 μm nominal diameter droplets. The profiles for two different 201 μm ± 1.34 μm droplets illuminated with 355 nm light are shown in figure 32, which are typical of these droplets. Note that the concentration of stilbene was increased to approximately 10 mg/L for the measurements of small droplets since the lower concentration of approximately 3 mg/L was not sufficient for imaging. The intensity profiles are qualitatively normal shaped. These droplets, however, are imaged with a 4 x 4 pixel array, which will lead to averaging of intensity over the droplet edges. For comparison, the predictions from Domann and Hardalupas[9] were pixelized using a 50 μm pixel size, and plotted using a smoothing function. This procedure was designed to estimate the intensity profile that would be detected by the camera. As with the 3 mm droplets, there is not satisfactory agreement between our measurements and the predictions of others for similar dye concentrations.

2.9 Droplet Sizing Method

The droplet sizes were determined by the following method:

1. First, verify the thickness of the measurement region. The intensities of 200 μm calibration droplet images were used to determine the background intensity and to separate droplets within the laser sheet to those outside the sheet. The 200 μm droplets were used for this determination since they produce the lowest intensity droplets within our measurement range. The width of the beam sheet was estimated from the imaged intensity profiles of the sheet projected on paper, as discussed previously. It was found that the signal intensities from the 200 μm and 3000 μm droplets fell sharply as they neared the edge of the sheet, at a location that coincided with the previous determination of the beam sheet edges. The intensities of the 200 μm droplets located just outside the sheet (+1 mm uncertainty) were taken as the intensity cutoff for droplets to be included in the measurements. Their intensity was determined to be approximately 20,000 units on a scale of 0 to 65535 for a 16 bit depth image using an HSI color model representation. (ie., $2^{16} = 65536$) This value was normalized to 0.305, allowing its application to other bit depths and images. It was assumed that the background was totally dark (I=0) which is a good assumption for these images. The resulting minimum detectable droplet image diameter is 166 μm, as computed from Equation 2 using the 0.305 cutoff intensity.

2. Next, apply a 2-D Sobel edge enhancing filter[19] to the image. The sobel filter replaces the intensities within the image with scaled values representing intensity gradients. Droplet images are converted to doughnut looking images after the application of the filter. It was found that the diameters of large and small droplets are accurately determined by measuring the diameter midway between the inner and outer edges of the doughnut at a normalized intensity of 0.305. Determining the droplet diameter by way of an intensity profile on an unfiltered image was found to cause greater errors on large and small droplet sizes than by using the sobel filtered images. The image dimensions were calibrated by photographing a steel ruler at the start of each roll of film.

Examples of small droplets, nominally 200 μm in diameter, are shown in figure 33. These droplets are shown in their unfiltered form and after the application of the sobel filter. Figure 34 and figure 35 show the normalized intensities of the images before and after the application of the sobel filter. These profiles were measured along a horizontal line cutting through the center of the droplet. The plots illustrate that it is difficult to determine the diameter of the drops by looking at the droplet intensity alone. To account for the eccentricity of the droplet images, the areas of the inner and outer donuts are used to calculate an equivalent diameter for each droplet. By using the donut method on the sobel filtered data, the diameter can be measured to an uncertainty within one pixel width. This will be demonstrated later.

Similar examples of large droplets are shown in figure 36. This figure shows a single droplet of approximately 3 mm diameter falling through the air. The images are presented before and after the application of the sobel filter. In this case the donut can be seen more clearly due to the larger number of pixels covering the drops. Notice that the droplet is not perfectly spherical in nature. The algorithm uses effective diameters based on the areas of the droplet and filtered images, which avoids an orientational bias in two dimensions. Plots of the droplet image intensity profiles are shown in figure 37 and figure 38. As in the case of the small droplets, the diameters are measured using the dimensions of the donut.

Using the above methodology, images containing calibration droplets were analyzed. Due to the slight curvature of the vertical iso-intensity profile of the laser sheet, the images of the droplet stream located very close to the edge of the sheet contain less droplets than those near the center. Images from droplets close to the sheet edge are also susceptible to small positional errors, since small changes in horizontal position are associated with steep changes in laser sheet intensity. The fraction of droplets measured by the analysis algorithm at various sheet positions is shown in figure 39. As discussed previously, despite some intensity profile curvature near the edge of the laser sheet, the droplet image drop off is steep at the edge of the sheet thickness. In addition, the plot shows similar results for analyses using 8 bit and 16 bit pixel depth. The use of 8 bit pixel depth cuts the size of the raw image files substantially. Note that the droplet fraction indicates a difference in laser sheet thickness for the small versus large drops. This is due to the use of the 8 mm slit in the large drop measurements, which reduced the thickness of the laser sheet by approximately 3.5 mm.

Due to the smaller number of droplets that will be imaged by the edge of the sheet, the statistics used to determine the uncertainty of the droplet size measurement are based on the total number of calibration droplets within the sheet. Therefore, the average droplet size at any particular location in the sheet is the average of the droplets measured at that horizontal position, but the droplet size average within the sheet is the average of all droplets images within the sheet, not the average of the average values determined at each horizontal location. The results of the calibration are summarized in Table 7, and illustrated in figure 40 through figure 43.

Note that the results of the calibration droplets agree with the fluorimeter predictions, where the signal level observed for the stilbene fluorescence is greater than the rhodamine fluorescence. For the determination of water flux and droplet size statistics, the blue droplet images are used due to the higher signal level image quality.

3. Measurements in Sprinkler Sprays

3.1 Sprinkler Spray Images

3.1.1 Fluorescence/Scattering Method

Images of fluorescence and Mie scattering from sprinkler sprays are shown in figure 44 and figure 45. The images are 84 mm by 118 mm sections of the overall imaged area of 25 mm by 35 mm. The f# was 2.8 and the magnification was 0.1. Note that the 355 nm excitation consistently gives a much redder drop than the 532 nm excitation, due to the scattered light. The 355 nm beam precedes the visible by 1 ms, which was verified using an oscilloscope triggered by a photodiode. The 532 nm laser power in figure 44 is 10 mJ/pulse while the power in figure 45 is 200 mJ/pulse. The 355 nm power (120 mJ/pulse) and the dye concentration are held constant.

At low 532 nm laser power, there are clear particle pairs throughout the image. The drops illuminated by the green laser are only slightly more yellow than the red drops, making distinction in colors difficult. Only due to the direction of gravity can the drop sequence be known and the velocity determined. The velocity of the large drops can be easily measured, but almost no small drops are visible. Increasing the 532 nm laser power to 200 mJ/pulse improves the color differentiation between the two drops. More particle pairs are indicated but unpaired green drops appear as a result of the higher intensity of scattering. The drop size at low 532 nm power was the same for both laser beams, while at higher laser powers all 355 nm drops are smaller than the corresponding 532 nm drops. Drops illuminated by 532 nm tend to have three types of images. In the first type, drops are solid or faint green with no fluorescence, indicating that they are at the edges of the beam or caused by secondary scattering of the laser sheet. The second type has a halo of green scattering completely or partially surrounding a fluorescent core. These are most likely small drops inside the light sheet or large drops partially in the light sheet. Finally, large drops have bright green scattering on the incident side of the laser beam while the rest of the drop image is dominated by a yellowish fluorescence that delineates the drop perimeter.

Figure 46 is an enlargement of a pair of drops in figure 45. In this image, two drops appear to coalesce as they fall. The initial UV beam indicates two fluorescence centers that are disconnected while the second visible laser indicates two bright centers with a single surrounding drop that is oblong. The upper most UV illuminated drop measures 1.47 mm or 21 pixels across. The velocity of the drops is easily measured from drop center to drop center. The radial component is 3.3 m/s outward and the axial component is 5.74 m/s downward, resulting in a speed of 6.6 m/s.

3.1.2 Two-Color Fluorescence Method

Images from the two-color fluorescence method are shown for two sprinklers in figure 47 and figure 48. The sheet breakup can be clearly seen in the 6 mm orifice case, and it occurs approximately xxx m away from the strike plate. The breakup in the 4 mm orifice case occurs closer to the strike plate, and continues as the ligaments also break up into droplets. In order to optimize the data gathered for each sprinkler, the measurement region is shifted radially and axially from the strike plate in order to allow the droplet density to decrease, and so that primary breakup occurs prior to the water entering the field of view of the camera. The measurement regions used for analysis are shown in figure 49 and figure 50. Magnified sections of the measurement regions are shown in figure 51 and figure 52. Droplet pairs of various sizes can be clearly seen in these photos.

3.2 Size and Velocity Distributions

3.2.1 Fluorescence/Scattering Method

Using the technique described above, images of sprinkler sprays from a nominally 8 mm nozzle at three combinations of flow rates and strike plate cone angles were analyzed. Fluorescence data was obtained using a Quantaray red pass filter to remove scattering. Thus, directional ambiguity was resolved only by the direction of gravity. An f# of 1.2 was used and the magnification was 0.1. As discussed earlier, the minimum fluorescence signal depends on the f# of the lens and the magnification. Based on Equation 1 and figure 17, the minimum detectable fluorescence results from a drop of approximately 35 μm. Thus the image resolution was dominated by the film resolution resulting in an expected minimum measurable diameter of 158 μm as determined from photographs of the resolution chart. Based on Kadambi's [7] criterion of 3 pixels per particle, the minimum measurable diameter is 210 μm. Although drops of smaller diameters can be detected, their diameters can not be accurately determined.

Measured particle trajectories are shown in Figure 53 as indicated by the instantaneous drop locations for the three cone angles. The sprays originated at the 0,0 point and were directed downward and outward at an angle determined by the cone angle of the nozzle. Measurements were made with 5 gpm and a cone angle of 90°, 4 gpm and 120° and 3 gpm and 140°. The corresponding velocities of particle pairs are shown in Figure 54. Very few images are required to determine the flow trajectories, velocities, and statistics of the spray distribution. For example, in this case two images were used.

Figure 55 through figure 57 describe the spray statistics. In each of these figures, the drop sizes below 150mm have been binned together at 100mm. The lowest resolvable drop size is 200mm, which is the second bin in the graphs. Figure 55 is the drop size distribution function. The peaks have been normalized to indicate the spread. The 5 gpm and 4 gpm drop distributions are similar, while 3 gpm indicates larger diameter drops that account for a higher percentage of flow. It should be noted again that although the drops cannot be more accurately sized, they are observable and therefore the large size bin into which they are placed indicates the possible error in size estimation.

Figure 56 is the cumulative number distribution, where 0.5 represents half of the total particles in the flow. The statistics indicate that over half of the observed particles in the flow are below the accurately measurable limit of the imaging systems. However, the cumulative volume fraction in Figure 57 indicates that these particles carry about

0.2% of the water flow rate. The cumulative number distribution also indicates that the lower flow rates have a higher fraction of larger drops, which is indicated in the cumulative volume fraction graph for 3 gpm.

3.2.2 Two-Color Fluorescence Method

The laser and imaging components for the sprinkler measurements have the same setup as used to gather the calibration images discussed earlier. The droplet size and velocity algorithms tested with the calibration drops are applied to the sprinkler images. Recall that the calibration process served to validate the measurement method, not to derive a correction factor since it was found that the droplet diameters were measured to within 6% of the actual drop size, with the uncertainty in the measurements encompassing the actual drop size value.

The drop size number fraction is shown in figure 58 for two sprinkler operating conditions: 90 degree strike plate, 6 gpm flow rate, and orifice diameters of 4 mm and 6 mm. The time delays between the pulses were set at 130 μs for the 4 mm orifice and 312 μs for the 6 mm orifice. The median droplet size based on count is approximately 220 μm for both sprinkler operating conditions, but 4.6% of the 6mm orifice drops and 41% of the 4 mm orifice drops were less than 200 μm design size. In order to better understand the droplet size distribution, the drop size number fraction is plotted in a logarithmic format in figure 59. If the drop size distribution is log-normal, which is useful for characterizing many spray nozzles, the distribution would be symmetric on the plot. The plot illustrates that the diameter of peak number fraction has been captured, but that many droplets under 200 μm in size have probably been missed by the measurement technique. This is not surprising since at the onset of this study the decision was made to capture a range of droplets larger than 200 μm, which in the past was found to contribute most to the overall water volume.

The droplet volume distribution is graphed in the same manner using logarithmic plots, shown in figure 60. If the drop distribution is log-normal in nature, then the drop volume distribution should also be symmetric, and of the same shape as the number distribution, but shifted to the right. In this case, it appears that the distribution does not fit the distribution well, especially for the 6 mm orifice size. The distribution does seem to indicate, however, that the measurements did a good job of capturing the droplets containing most of the water from the sprinkler. This is also illustrated by examining the cumulative water volume fraction shown in figure 61. This graph shows that the original premise of the measurement design holds true, that droplets greater than 200 μm in diameter contain more than 95% of the water ejected by the sprinkler. (98.7% for the 6 mm orifice, and 95.1% for the 4 mm orifice.) This condition holds true for sprinklers producing a wide range of droplet sizes and operating conditions. The 4 mm sprinkler produces 50% of the water volume at drop sizes greater than approximately 450 μm, while the 6 mm sprinkler produces 50% of the water volume at drop sizes greater than approximately 1140 μm. Note that these two sprinklers have nearly identical number median drop sizes.

The above results were derived from averaging the results of 8 film exposures. One of the questions to be answered is how many images are necessary to suitably characterize the flow from the sprinkler. For the axis-symmetric sprinklers studied here, the results are summarized in figure 62 and figure 63. In these figures, the results from one exposure and from averaging the results from 4 and 8 exposures are plotted for comparison. For the 4 mm orifice sprinkler, the results from 4 and 8 exposures is nearly identical, while the results from 1 exposure are all within approximately 10% of the 4 and 8 exposure values over the entire droplet diameter range. For the 6 mm orifice, the 4 and 8 exposure results were within approximately 2% of each other, while the 1 exposure result was within approximately 15% of the 4 and 8 exposure values over the entire droplet diameter range. These results indicate that for the sprinklers studied here, approximately 4 exposures would be sufficient for determining the droplet size statistics, and less than 4 exposures are acceptable if a higher level of uncertainty can be tolerated or if time varying behavior is to be captured.

Droplet velocities for one exposure of each sprinkler are shown in figure 64 and figure 65. Since the position and velocity of each droplet pair is available, the vector data was plotted on a single graph for each sprinkler. The plot is cluttered, but is included to show the density at which measurements were taken. In difference to PIV results, the velocities were not averaged and plotted for interrogation regions since this would disregard the velocity differences between droplet sizes within each region. The full data set is used in future investigations to validate trajectory predictions.

The average droplet speed for various droplet sizes is shown in figure 66. As expected, the magnitude of the droplet velocities are higher for the 4 mm diameter orifice since the initial velocity of the emerging water jet is higher than that of the 6 mm diameter orifice. Given that the emerging water jets are approximately 29 m/s and 13 m/s for the

4 mm and 6 mm orifices, respectively, the water has lost significant momentum due to drag, especially at smaller droplet sizes. The effects are greater for the droplets from the 4 mm orifice, where greater fraction of the breakup occurs closer to the strike plate, as opposed to the 6 mm orifice where the breakup occurs from a water sheet distant from the strike plate leaving less travel distance.

The deviation of the velocity vector from the angle of the strike plate is shown in figure 67. The data indicates that most of the water can be found within a small range of cone angles as can also be seen from figures 64 & 65. For the sprinklers investigated, most of the water was contained within a range defined by an upper bound defined by the strike plate angle, and the lower bound located at 10 degrees down toward the sprinkler axis of symmetry.

Finally as a check, the droplet velocity data is used together with the droplet sizing to check for mass conservation. The flow rate is calculated as the product of the water concentration within the sprinkler spray, the downward velocity of the sprinkler droplets, and the horizontal area of the spray as it exits the measurement region. This is expressed in Equation 11. The water flow rate computed from integrating the planar laser measurements matched the flow rate from the turbine flow meter within the flow meter uncertainty for both sprinkler operation conditions. For cases where the agreement was not as good, it was found that the sprinkler spray was not axis-symmetric – a condition implicitly assumed in Equation 11.

Equation 11

$$\dot{V} = \pi\left(r_2^2 - r_1^2\right) \sum_{i=d_{min}}^{d_{max}} \left(\frac{V_{water}}{V_{air}}\right)_i \left(v_{drop}\right)_i$$

Where: V = water volume flow rate
V_{water}=volume of water within measurement volume
V_{air}=volume of the measurement volume
v_{drop}=downward droplet velocity component
r_2=outer edge of measurement region
r_1=inner edge of measurement region
d_{min}=smallest drop diameter
d_{max}=largest drop diameter

4. Summary & Conclusions

In this work techniques were developed to make simultaneous particle tracking velocimetry and particle size measurements of water droplets in large-scale sprinkler flows. Spray patterns, drop velocities, and drop sizes were measured to provide data on instantaneous and time averaged water delivery density for five separate spray conditions. In addition, two sub-methods were presented: a scattering/fluorescence technique, and a two-color fluorescence technique.

In measuring large fields of view, detection of drop diameters down to 35μm is possible but size measurement is limited by the fluorescence intensity and collection optics parameters. In order to detect small drops, a low f# is desired, however there is limit to the f# beyond which the ability to accurately measure size is impeded. Reasonably accurate measurements of drop sizes greater than 200 μm were obtained in this work in fields of view of 0.25 m x 0.35 m and 0.46 m x 0.54 m by using laser-induced fluorescence. It should be noted that for smaller fields of view or higher magnification, proportionately smaller droplet sizes can be measured. The dye concentration in the water droplets was optimized to obtain maximum fluorescence intensity. This concentration was approximately 3.3 mg/L for rhodamine and 10. mg/L for stilbene. Only the film resolution and the standard digital resolution requirement of 3 to 4 pixels per particle limit these measurements. Scattering images did not prove to adequately represent the particle size. The effect of variation of intensity across the beam sheet, depth of focus, and volumetric response of the fluorescence signal combine to give varying drop sizes. With careful design of the experimental apparatus and the use of automated image analysis techniques, the size of droplets throughout the measurement volume were measured with uncertainties of 17% for 200 μm droplets and 7% for 3000 μm droplets. Sizing errors averaged from 2% for the 200 μm droplets to 6% for the 3000 μm droplets, with the measured size larger than the actual drop size in both cases.

Very few images are needed to characterize the spray, with 4 images appearing be the minimum necessary for determining the cumulative volume distribution to within 2% of larger numbers of exposures. Finally, spray number distributions indicate that many of the drops are below the measurable size, but volume measurements indicate that greater than 95% of the flow is carried by drops of measurable size. This conclusion is also supported by a mass balance, which indicated that the computed water flow volume from the images was within the experimental uncertainty of the water flow meters.

5. References

[1] McGrattan, K.B. Hamins, A. and Forney, G.P., 1999, "Modeling of Sprinkler, Vent and Draft Curtain Interaction," Sixth (International) Symposium on Fire Safety Science, France.

[2] McGrattan, K.B., Baum, H.R., Rehm, R.G., Hamins, A.H., Forney, G.P., Prasad, K., Floyd, J.E., and Hostikka, S. Fire Dynamics Simulator (Version 3) – Technical Reference Guide, NISTIR 6783, 2002 Ed., National Institute of Standards and Technology, Gaithersburg, MD, November 2002.

[3] W.D. Bachalo. Experimental Methods in Multiphase Flows. Int. J. Multiphase Flow. Vol. 20, Suppl. pp261-295. 1994.

[4] Widmann, J.F. Characterization of a Residential Fire Sprinkler Using Phase Doppler Interferometry. NISTIR 6561, National Institute of Standards and Technology, Gaithersburg, MD, August 2000.

[5] C.R. Tuck, M.C. Butler Ellis, and P.C.H. Miller. Techniques for the measurement of droplet size and velocity distributions in agriculture sprays. Crop Protection. Vol. 16 No. 7. pp619-628. 1997.

[6] Putorti, A.D., Belsinger, T.D., and Twilley, W.H. Determination of Water Spray Drop Size and Speed from a Standard Orifice Pendent Spray Sprinkler. NISTFR 4003, 6561, National Institute of Standards and Technology, Gaithersburg, MD, September 1995, revised May 1999.

[7] J.R. Kadambi, W.T. Martin, S. Amirthaganesh, and M.P. Wernet. Particle sizing using Particle Imaging Velocimetry for two-phase flows. Powder Technology Vol. 100. pp251- 259. 1998.

[8] D.C. Herpfer, S. Jeng. Planar Measurements of Droplet Velocities and Sizes Within a Simplex Atomizer. AIAA Journal, Vol. 35 No. 1. January. pp127-132. 1997.

[9] Domann, R. and Hardalupas, Y. Spatial distribution of fluorescence intensity within large droplets and its dependence on dye concentration. Applied Optics, Vol. 40, No. 21, 20 July 2001.

[10] Domann, R. and Hardalupas, Y. A study of parameters that influence the accuracy of the planar droplet sizing (PDS) technique. Part. Syst. Charact. 18(2001) 3-11.

[11] Z. Cao, K. Nishino, and K. Torii. Measurement of Size and Velocity of Water Spray Particle Uising Laser-Induced Fluorescence Method. Proceedings of the 2nd Pacific Symposium on Flow Visualization. Honolulu, Hi. May 16-19, 1999.

[12] Adrian, R., 1991, "Particle-Imaging Techniques for Experimental Fluid Mechanics," Annu. Rev. Fluid Mech., Vol 23, pp 261-304.

[13] Modern Optical Engineering. W.J. Smith. McGraw-Hill, New York, third ed., 2000.

[14] Introduction to the Optical Transfer Function. C.S. Williams and O.A. Becklund. John Wiler & Sons, New York, 1989.

[15] Star Testing Astronomical Telescopes: A Manual for Optical Evaluation and Adjustment. H.R. Suiter. Willmann-Bell, Richmond, VA, 1994.

[16] Fujicolor Portrait film, NPZ 800 Professional, Data Sheet, AF3-089E, Fuji Photo Film Co., Ltd., Tokyo, Japan.

[17] Photodo AB, Lund, Sweden. www.photodo.com

[18] Perception of Displayed Information. L.M. Biberman, ed. Plenum Press, New York, 1973.

[19] Image-Pro Plus Reference Guide, Media Cybernetics, Silver Spring, MD

[20] D. Everest and A. Atreya. Simultaneous Measurements of Drop Size and Velocity in Large-Scale Sprinkler Flows Using Laser-Induced Fluorescence. Proceedings of the 2nd Pacific Symposium on Flow Visualization. Honolulu, Hi. May 16-19, 1999.

FIGURES

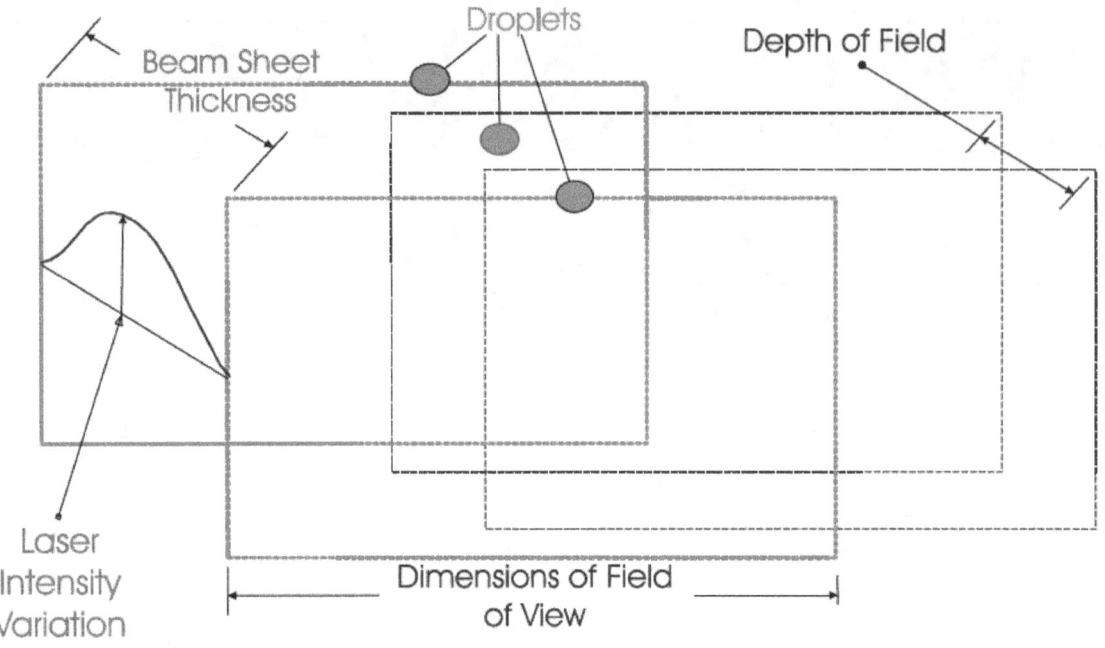

Magnification = Field of Image / Field of view

Figure 1. A schematic diagram showing the variables that must be considered for accurate particle size measurements by the Particle Tracking Velocimetry and Imaging (PTVI) technique.

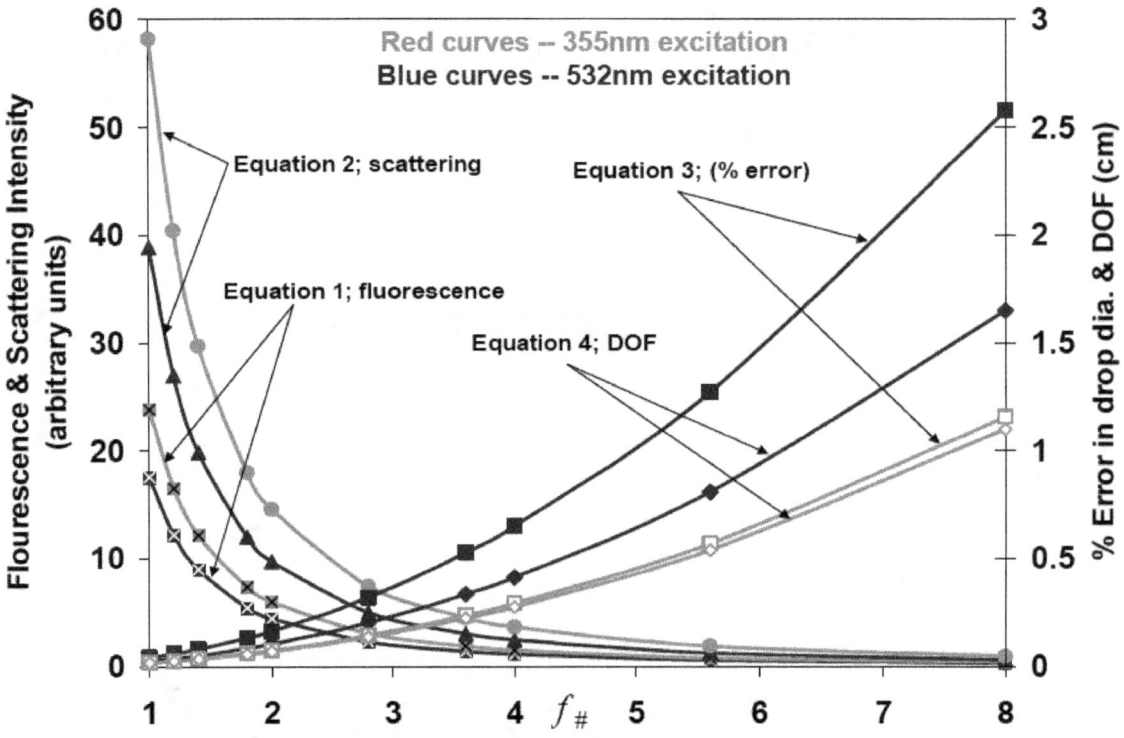

Figure 2. Variation of fluorescence and scattering intensity, percent diffraction error, and DOF (in cm) with f#. Red – 355 nm excitation, Blue – 532 nm excitation.

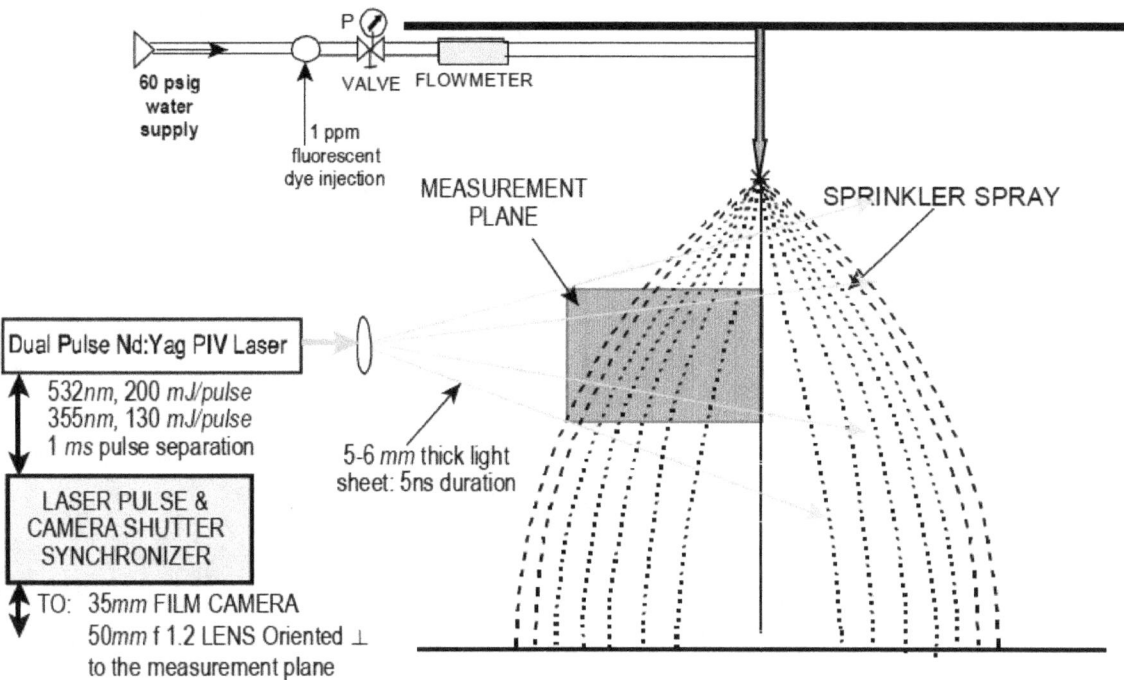

Figure 3:. Experimental arrangement for spray measurements.

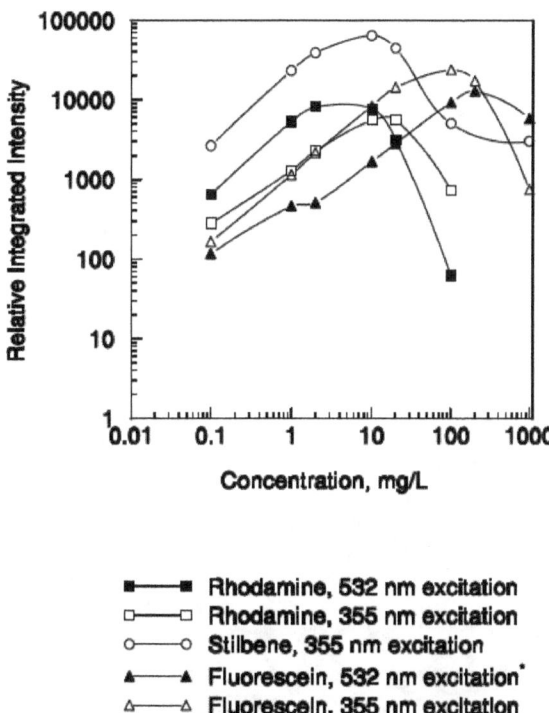

Figure 3. Relative fluorescent emission intensity from dye-water solutions integrated over all emission wavelengths.

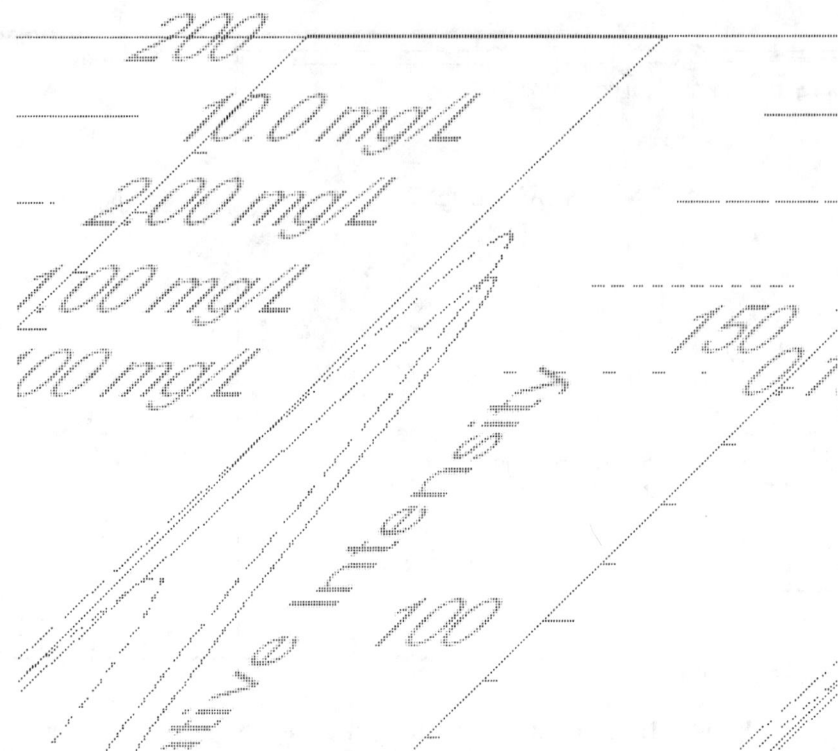

Figure 5. Relative fluorescent emission intensity spectra from rhodamine-water solutions excited by 532 nm light. The response maximum for 2.00 mg/L of dye is at 584.5 nm.

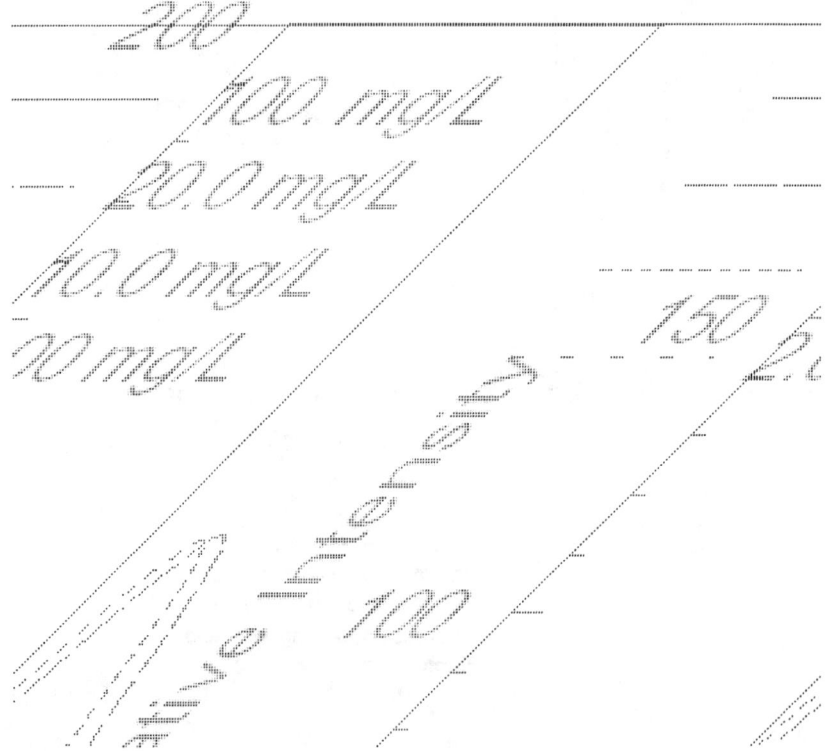

Figure 6. Relative fluorescent emission intensity spectra from rhodamine-water solutions excited by 355 nm light. The response maximum for 20.0 mg/L of dye is at 594 nm.

Figure 7. Relative fluorescent emission intensity spectra from stilbene-water solutions excited by 355 nm light. The response maximum for 9.96 mg/L of dye is at 434 nm.

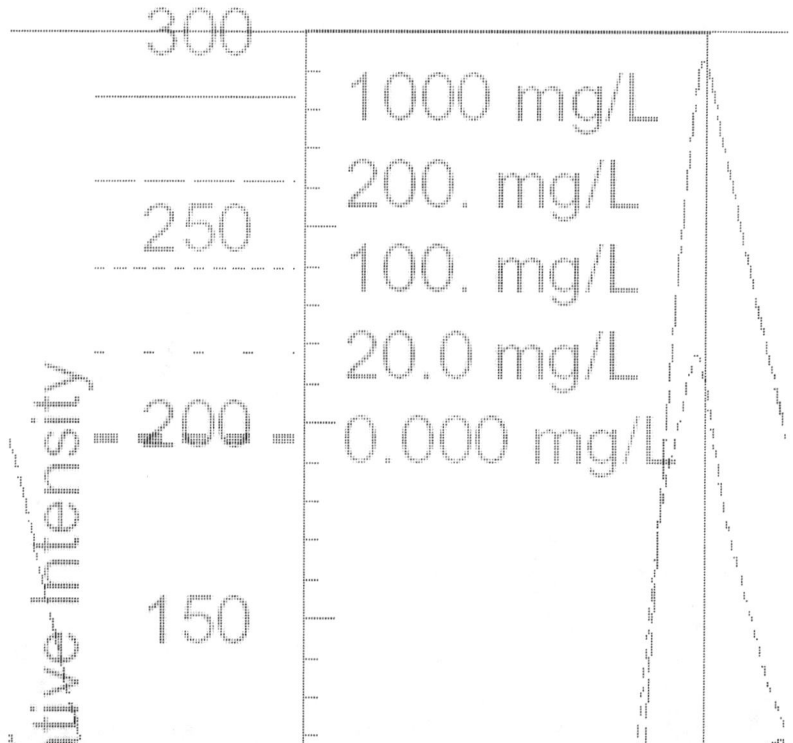

Figure 8. Relative fluorescent emission intensity spectra from fluorescein-water solutions excited by 532 nm light. The response maximum for 200. mg/L of dye is at 534.5 nm. Note that fluorescein may not be fluorescing at this wavelength due to uncertainty in the incident wavelength, but the response curve does indicate that it is fluorescing about the excitation wavelength of 532 nm.

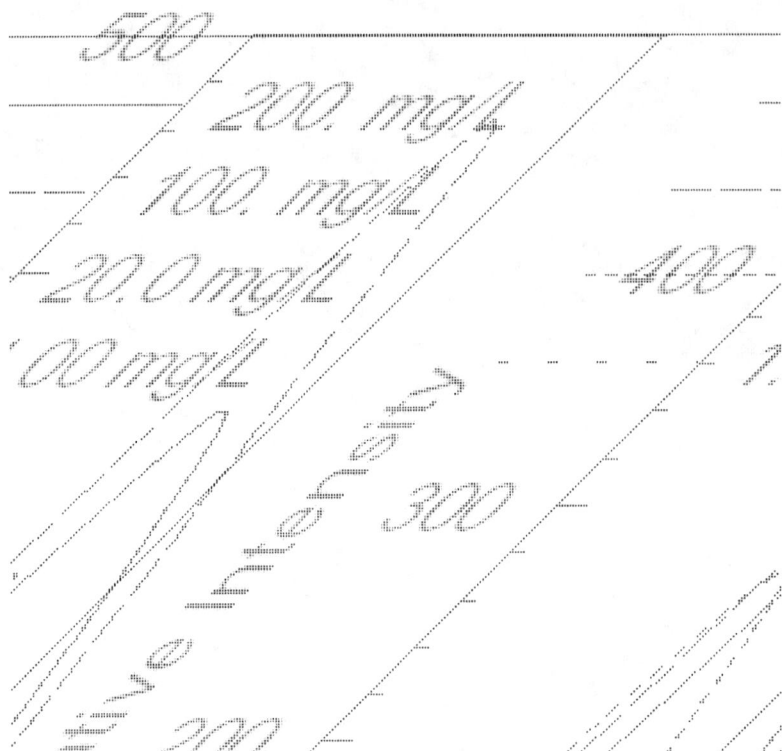

Figure 9. Relative fluorescent emission intensity spectra from fluorescein-water solutions excited by 355 nm light. The response maximum for 100. mg/L of dye is at 532.5 nm.

Figure 10. Relative fluorescent emission intensity spectra from rhodamine-stilbene-water solutions excited by 355 nm light. The response maximum for 3.33 mg/L rhodamine and 9.65 mg/L stilbene in water is at 433 nm.

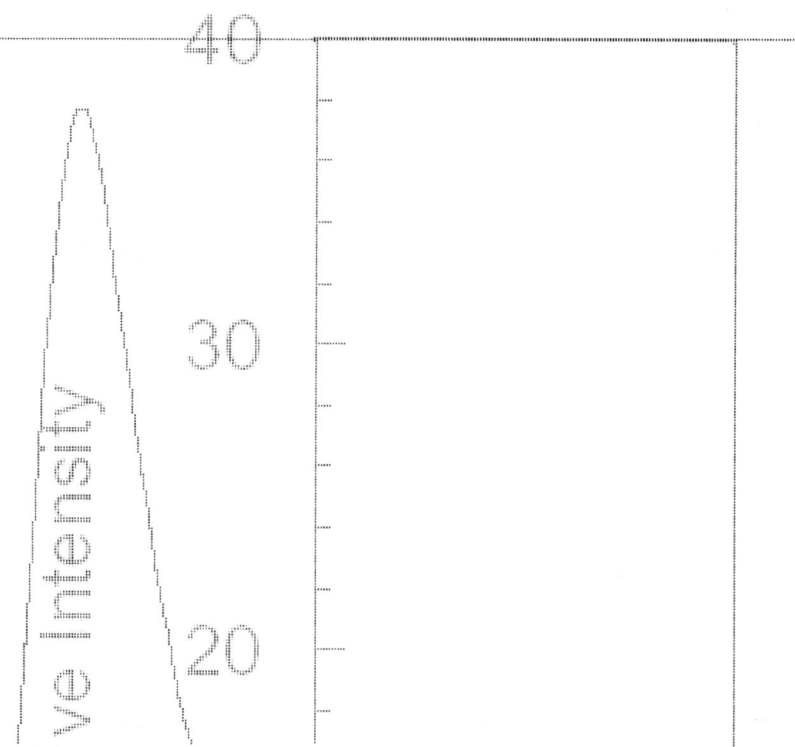

Figure 11. Relative fluorescent emission intensity spectra from rhodamine-stilbene-water solution excited by 434 nm light, which is the wavelength of the fluorescence emission maximum for stilbene-water mixtures excited by 355 nm. The rhodamine in the mixture is absorbing some of the 434 nm light, and fluorescing at 590 nm.

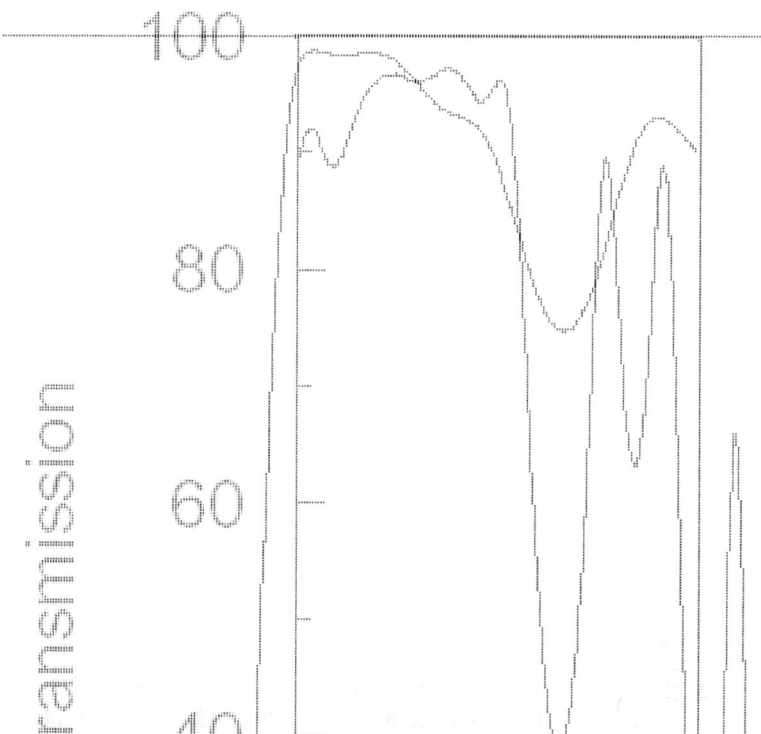

Figure 12. Transmission function for 532 nm notch filter. The filter is used on the camera during the two-color fluorescence measurements to exclude 532 nm light that will be scattered by the water droplets.

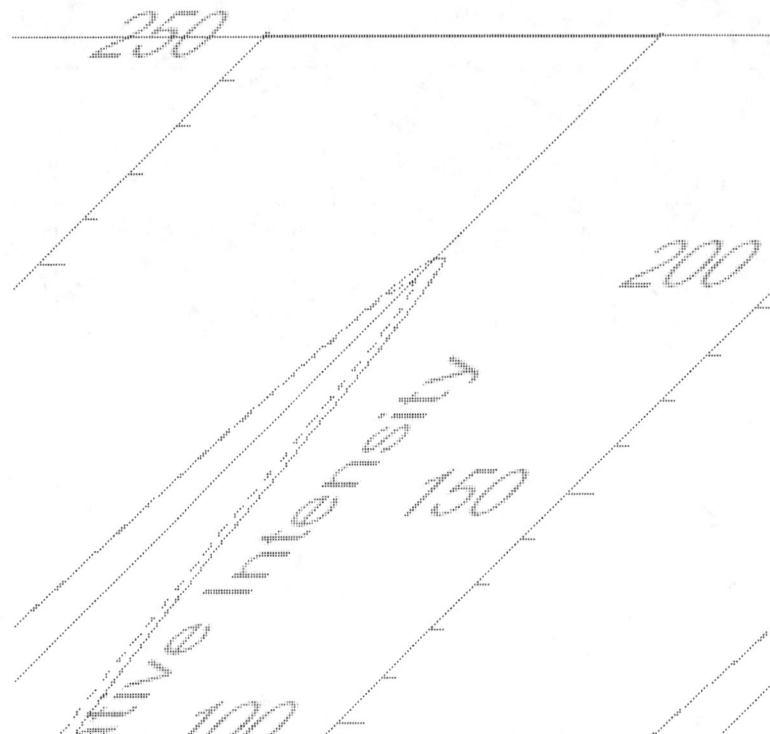

Figure 13. Fluorescent emission intensity spectra of rhodamine-stilbene-water solutions excited by 532 nm light. Comparison of results with and without 532 nm notch filter. The peak emission wavelength without the filter is at 587 nm, and the peak emission wavelength with the filter is at 586 nm. With the filter installed, 50% of the fluorescent light intensity is above 592 nm. This is one of the wavelengths used for the DOF calculations.

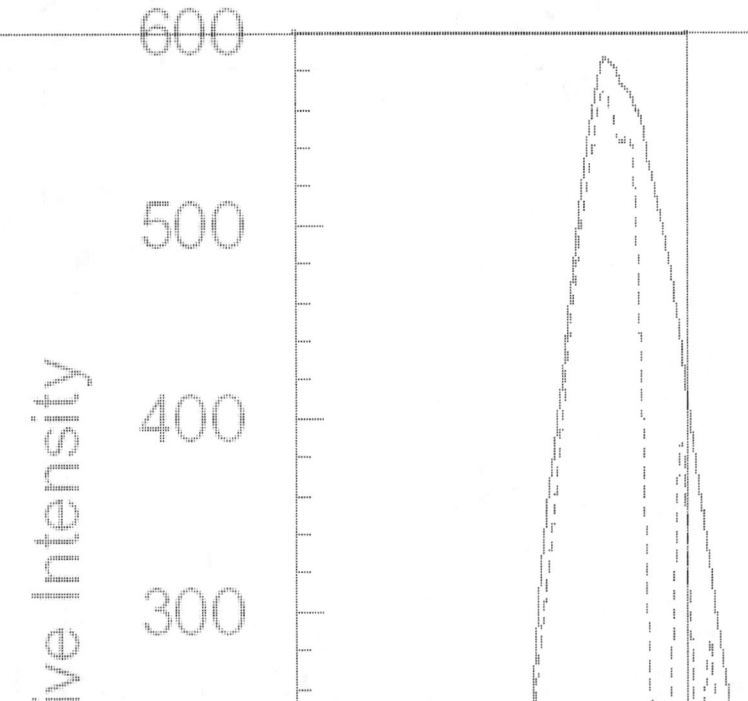

Figure 14. Fluorescent emission intensity spectra of rhodamine-stilbene-water solutions excited by 355 nm light. Comparison of results with and without 532 nm notch filter. The peak emission wavelengths with and without the filter are at 433 nm. With the filter installed, 50% of the fluorescent light intensity is above 439 nm. This is one of the wavelengths used for the DOF calculations.

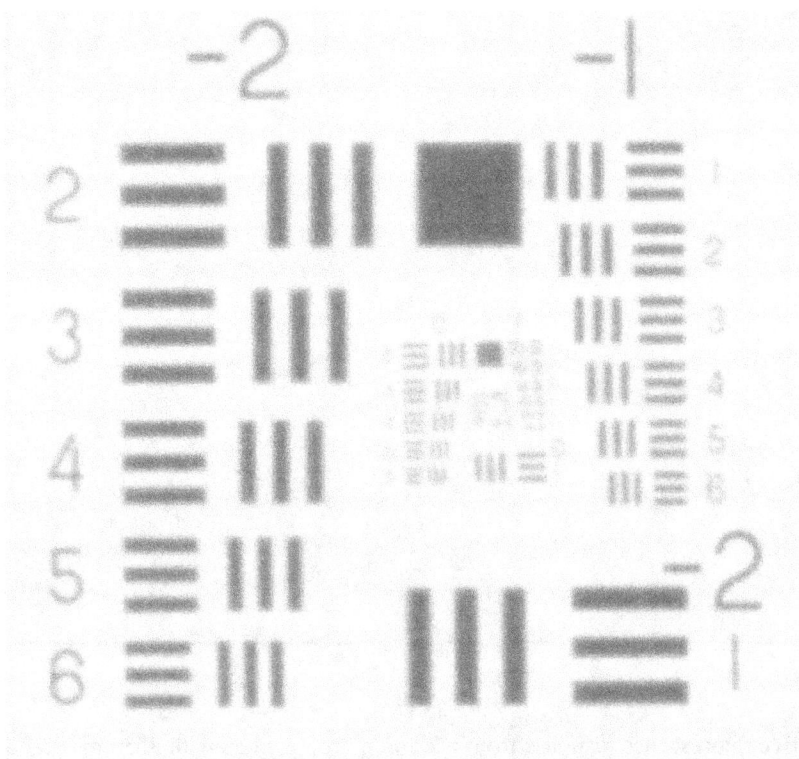

Figure 15. A portion of the USAF 1951 resolution chart. The chart is imaged and used to determine the resolution of the imaging system.

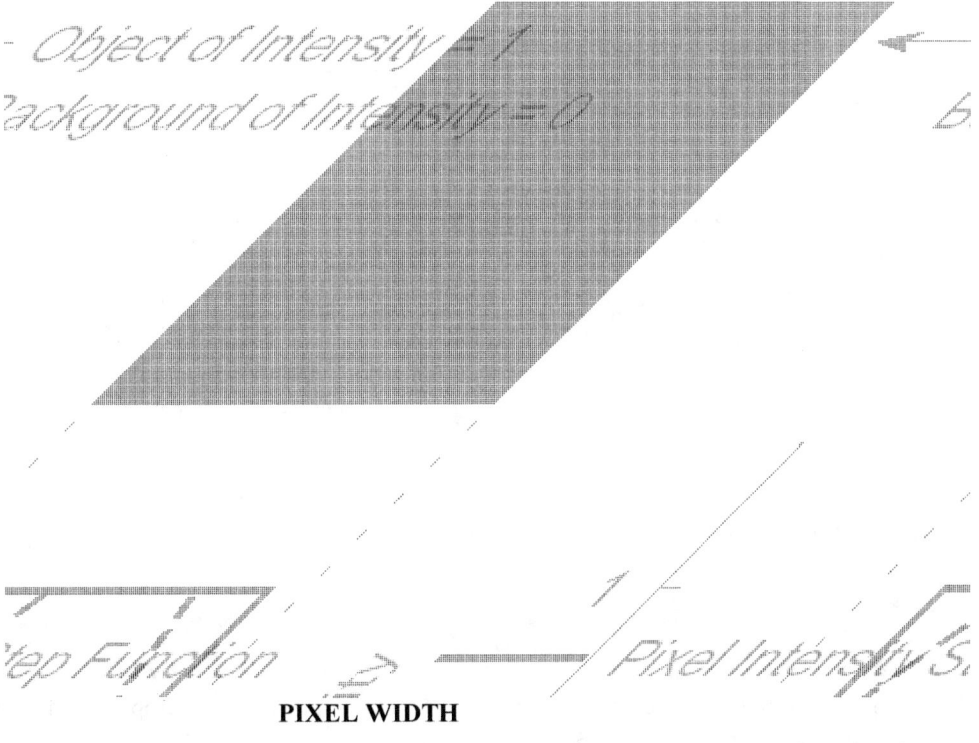

PIXEL WIDTH

Figure 16. Pixel averaging effect on the edge of an object.

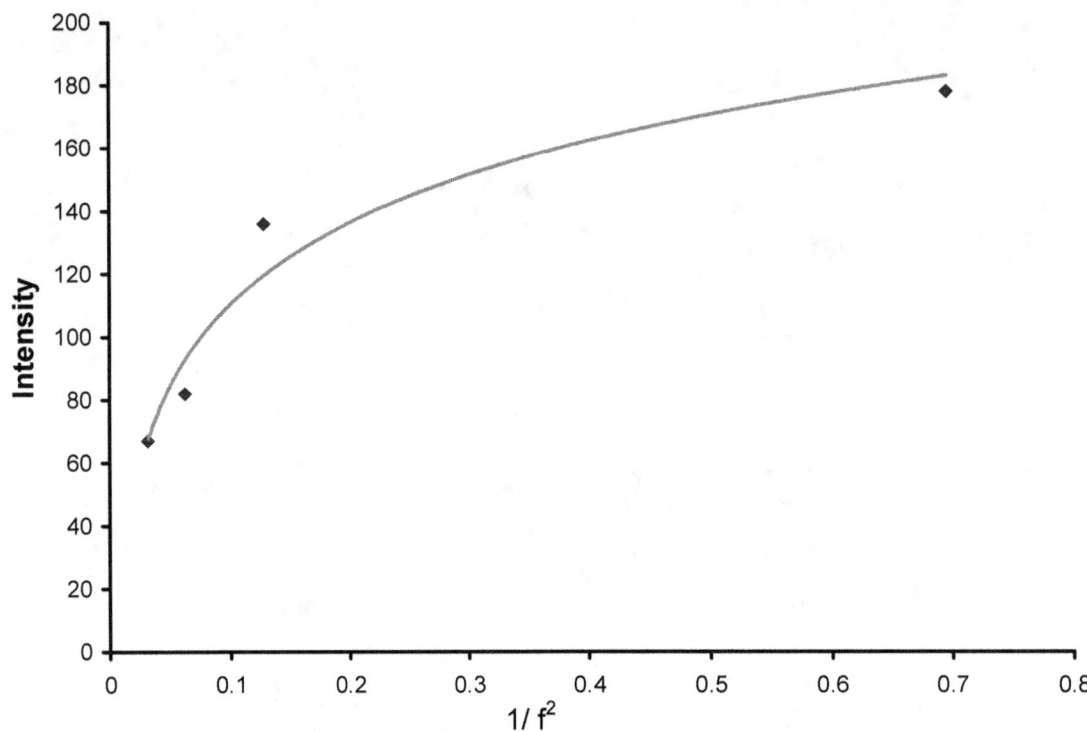

Figure 17. Relative fluorescence response from a 2.8 mm drop excited with 355 nm beam for a range of camera aperture f#s (arbitrary units).

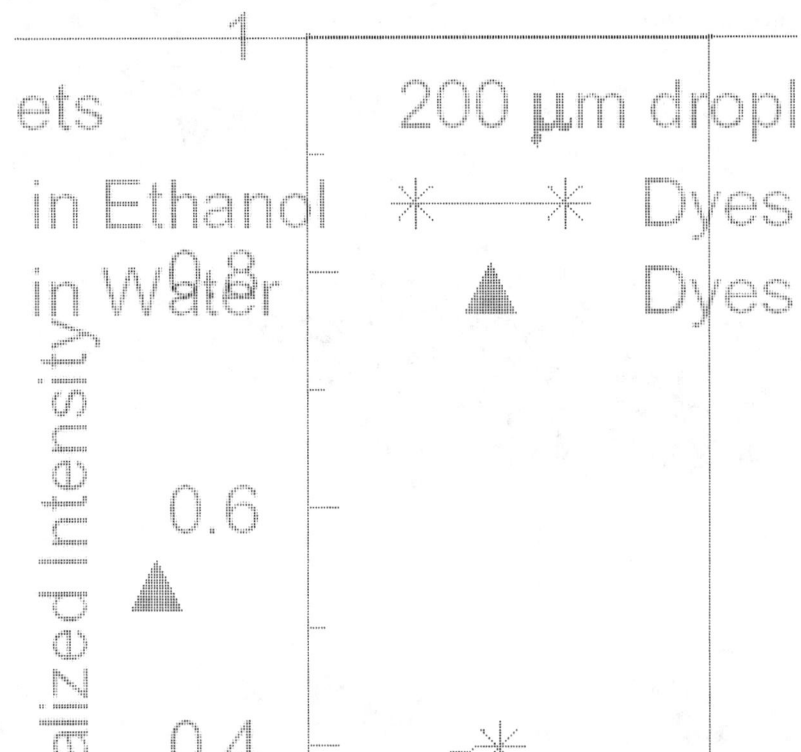

Figure 18. Normalized fluorescence response from a 200 □m nominal diameter drop excited with a 355 nm beam for a range of camera aperture f#s.

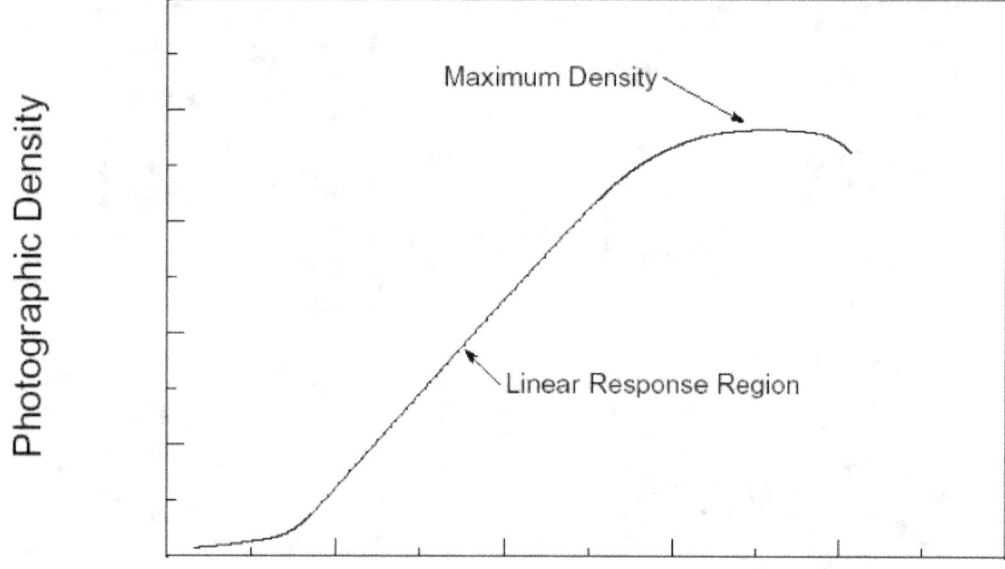

Figure 19. Response curve for a typical photographic film. Photographic density is measured by passing light through the negative, and is calculated as log10 (incident intensity/transmitted intensity).

Laser Beam Sheet Thickness

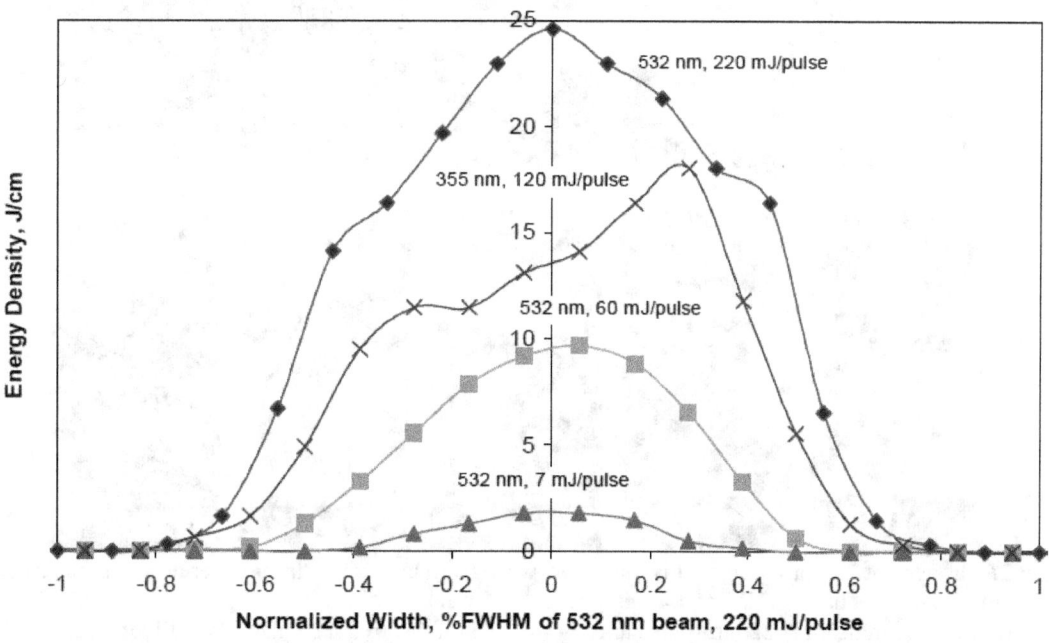

Figure 20. Relative beam sheet thickness for 355 nm laser beam at high laser intensity and for 3 different laser intensities of the 532 nm beam.

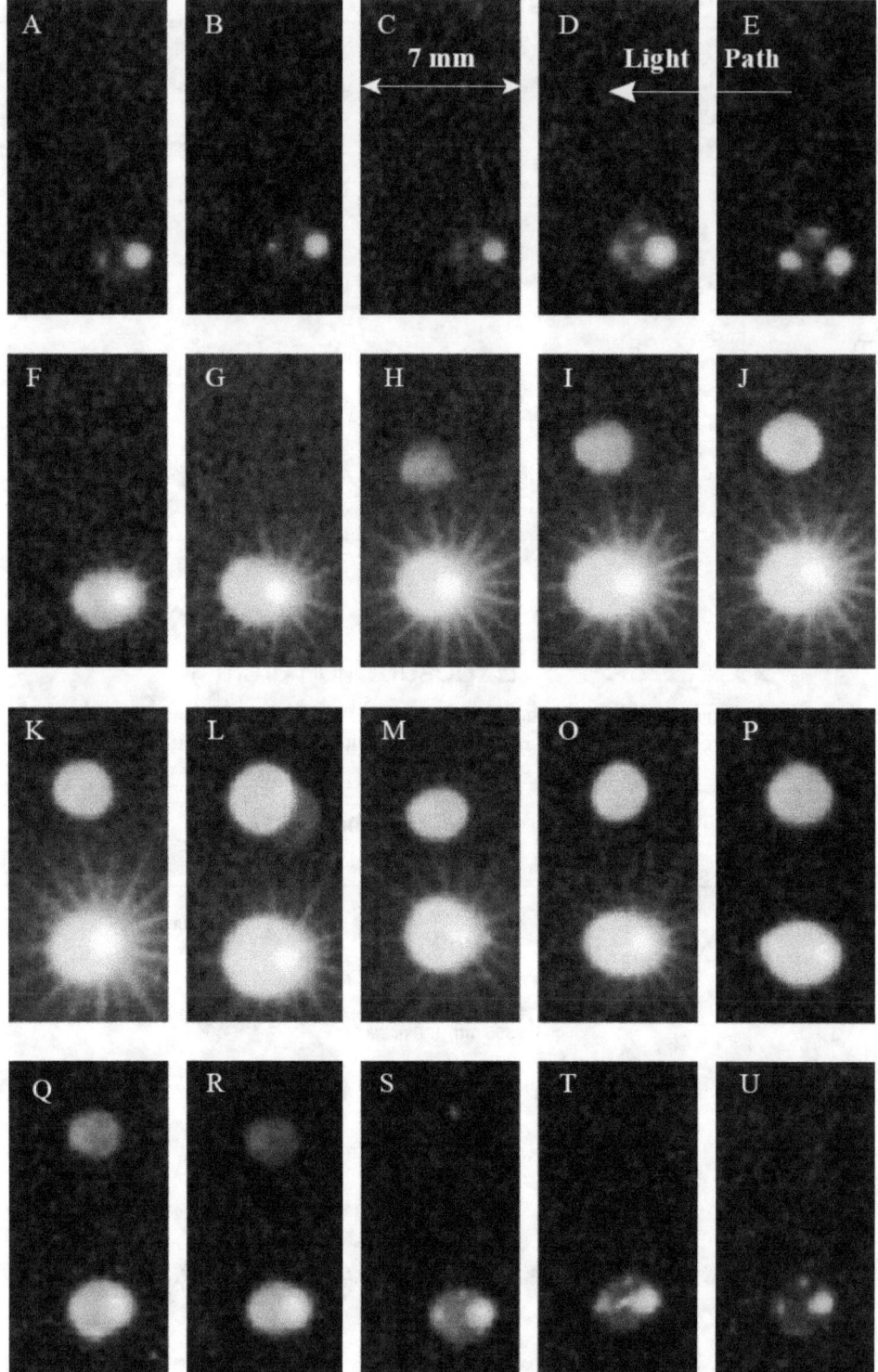

Figure 21. Images of 2.8 mm drop as it is moved forward from behind the laser sheet in 1 mm increments. The 532 nm laser sheet has FWHM of 4.4 mm and the 355 nm laser sheet has FWHM of 3.6 mm. Laser power for the 532 nm beam is 220 mJ/pulse and for the 355 nm beam is 130 mJ/pulse. Fluorescence at 588 nm from 355 nm excitation is observed from the upper drop in each frame, while scattering and fluorescence from 532 nm excitation is observed from the lower drop (Filters were not used in taking these images). When the drop is centered in the beam sheets, the drop size as measured by the 355 nm excitation is 2.7 mm, while the drop size as measured by 532 nm excitation and scattering is 3.3 mm.

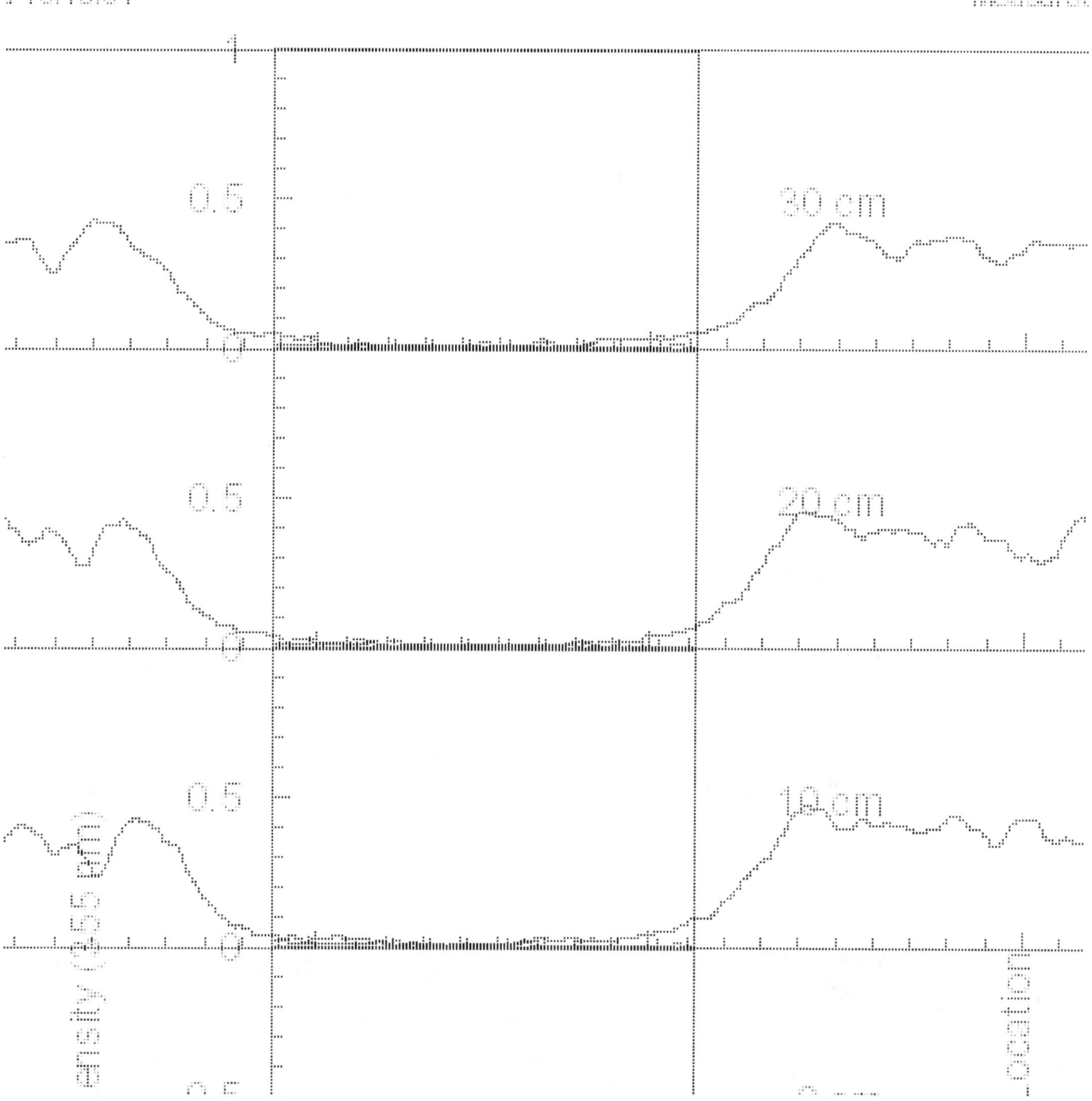

Figure 22. Laser sheet power profile measured at the near edge of the droplet measurement region. Laser power is 4.78W at 355nm. Beam sheet thickness is approximately 13.5mm over the vertical range of 0.6 m.

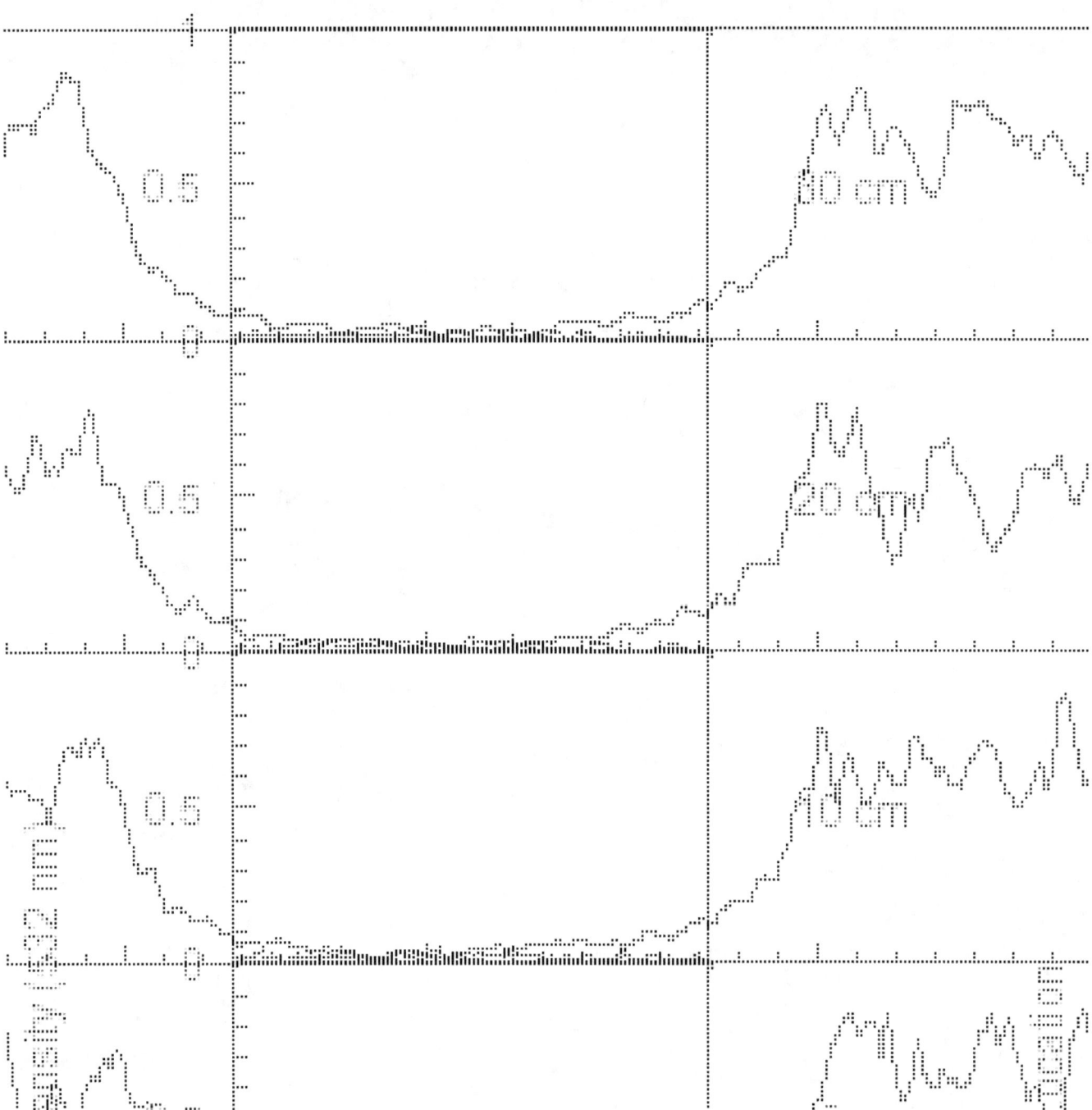

Figure 23. Laser sheet power profile measured at the near edge of the droplet measurement region. Laser power is 5.60W at 532nm. Beam sheet thickness is approximately 10.0mm over the vertical range of 0.6 m.

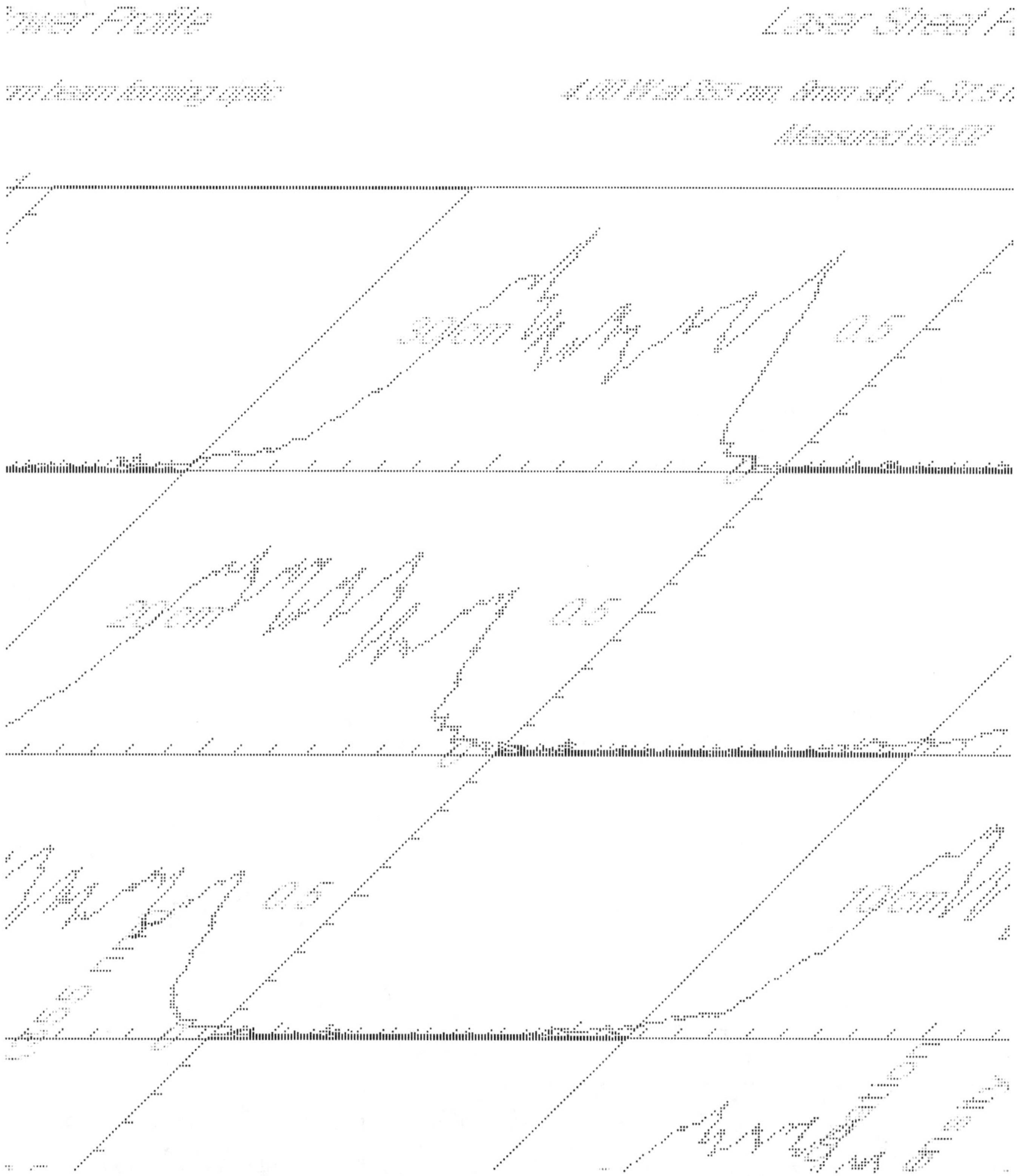

Figure 24. Laser sheet power profile measured at the near edge of the droplet measurement region. Laser power is 4.00 W at 355 nm. 8 mm slit is installed after sheet forming optic. Beam sheet thickness is approximately 10.5 mm over the vertical range of 0.6 m.

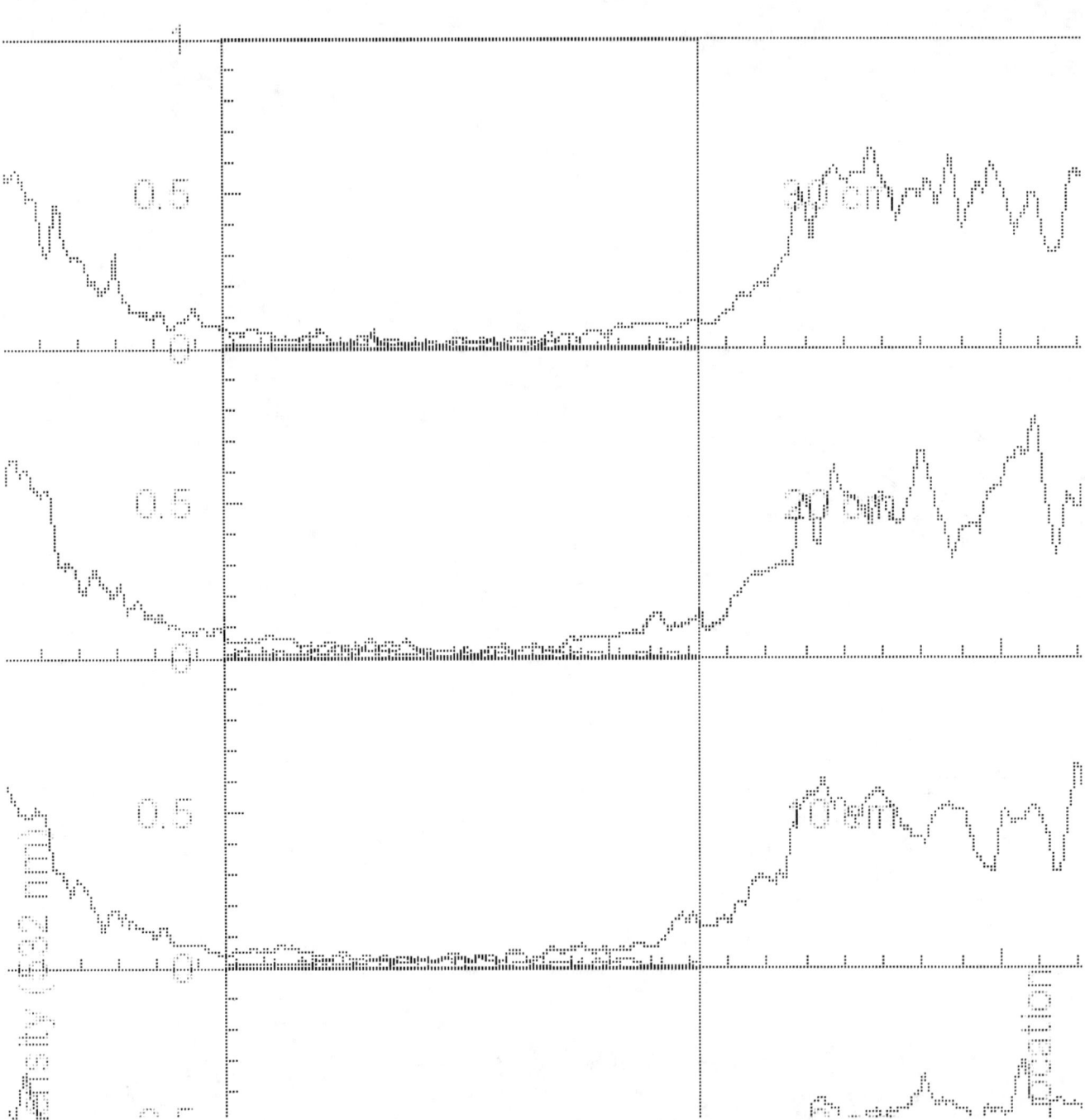

Figure 25. Laser sheet power profile measured at the near edge of the droplet measurement region. Laser power is 7.20 W at 532 nm. 8 mm slit is installed after sheet forming optic. Beam sheet thickness is approximately 10.0 mm over the vertical range of 0.6 m.

Predicted Depth of Field

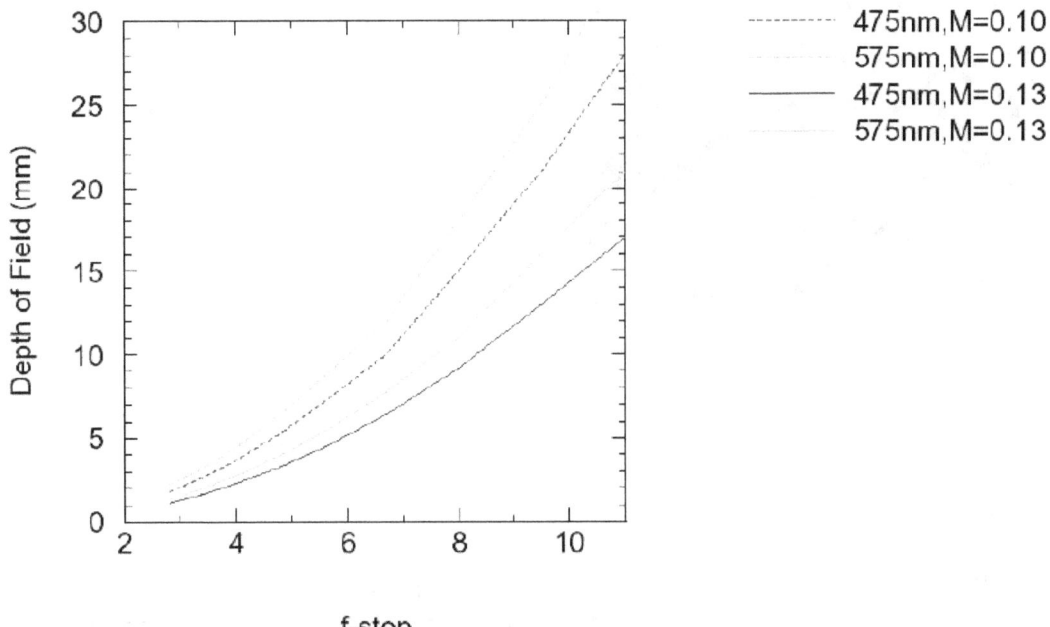

Figure 26. Predicted depth of field for various combinations of magnification, fluorescence emission wavelength, and f#. The plots are based on fluorescent emission wavelengths from a nominal solution of 3 mg/L of rhodamine and 10 mg/L of stilbene in water using the 532 nm notch filter. Note that 439 nm is the 50% cumulative emission intensity wavelength for 355 nm excitation, and 592 nm is the 50% cumulative emission intensity wavelength for 532 nm excitation.

A → f=2.8   B → f=4.0   C → f=5.6   D → f=8.0

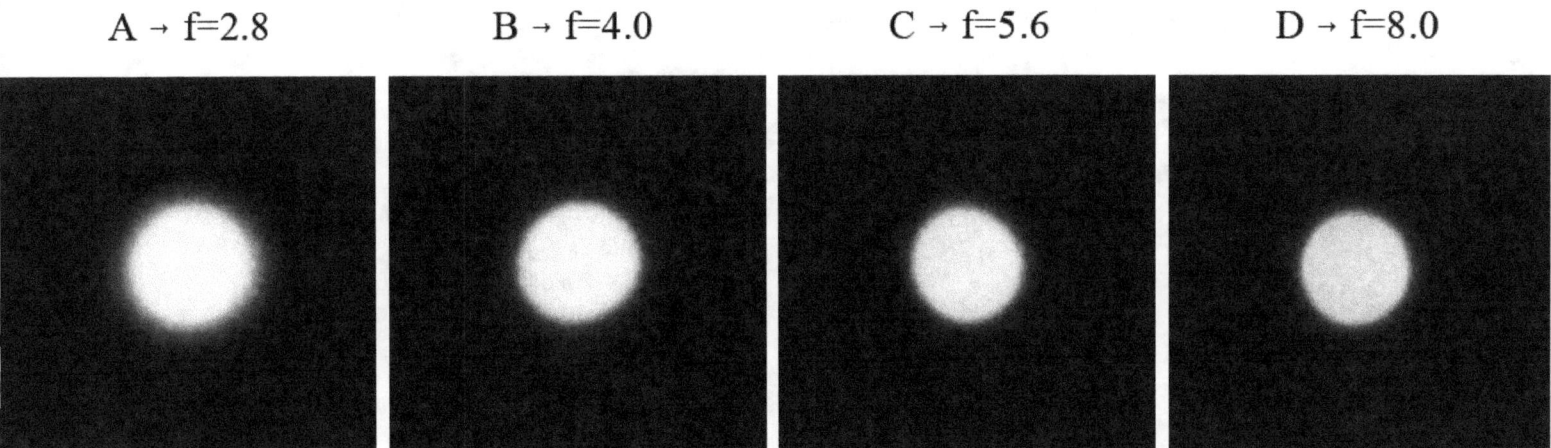

Figure 27. Effect of f# on droplet images. 2800 μm diameter water drops with approximately 3 mg/L of stilbene and 3.3 mg/L of rhodamine illuminated with 355 nm laser sheet. Image quality improves as f# is increased from 2.8 to 8.0 as evidenced by haze around droplet perimeter. Images taken at f#s of 5.6 and 8.0 are of nearly equal quality.

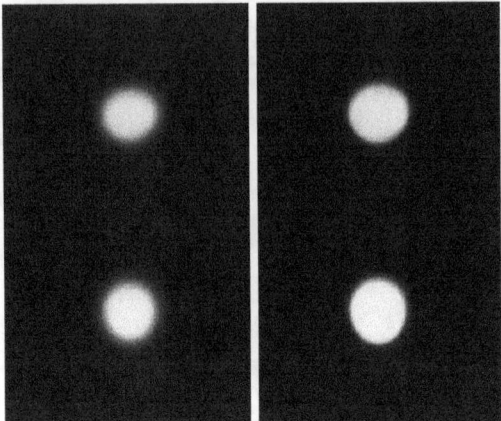

Figure 28. Effect of scanner focus on droplet images. The images in this figure show nominally 3000 μm diameter droplets illuminated with 532 nm (upper drops) and 355 nm (lower drops) laser sheets. The images are two successive scans of the same film. Care must be taken to verify that the scanner is properly focused, a procedure that is important with small pixel sizes.

R6E1 large droplet profile

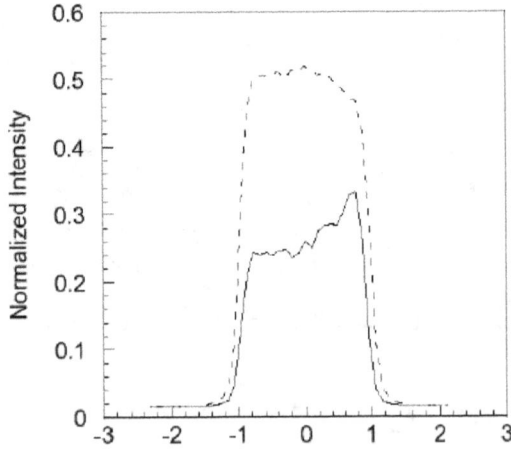

——— Blue (Stilbene) Drop
- - - - Yellow (Rhodamine) Drop

Figure 29. Horizontal intensity profile of 3 mm nominal diameter water droplet conducted through the centerline. Droplet was illuminated with 532 nm and 355 nm laser sheets. The intensity is normalized by the maximum scanner intensity, and plotted against distance, normalized by the blue drop image radius.

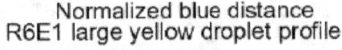

R6E1 large yellow droplet profile

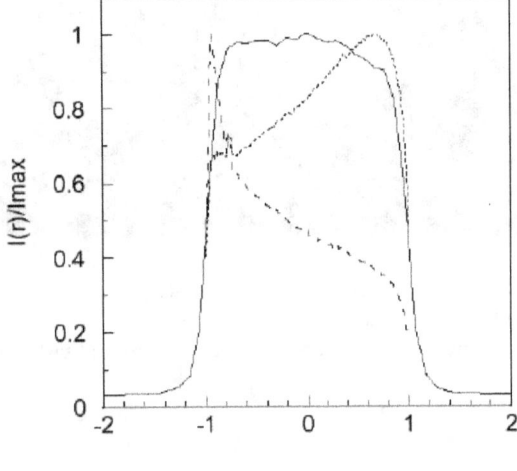

——— 3.3 mg/L Rhodamine WT
$(5.7 \times 10^{-6}$ M) Data
- - - ≈10 mg/L Rhodamine 6G
$(2 \times 10^{-5}$ M) Prediction
········· ≈500 mg/L Rhodamine 6G
$(1 \times 10^{-3}$ M) Prediction

Figure 30. Horizontal intensity profile through the centerline of a 3 mm nominal diameter water drop illuminated by 532 nm light compared to profiles predicted by others. Intensity is normalized by the maximum droplet intensity, and the horizontal dimension is normalized by the droplet radius.

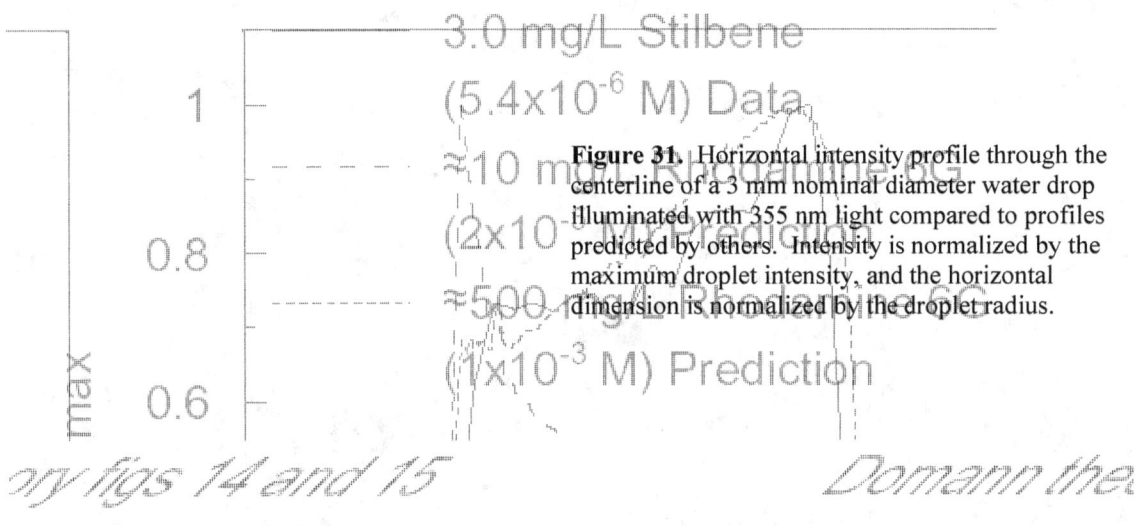

Figure 31. Horizontal intensity profile through the centerline of a 3 mm nominal diameter water drop illuminated with 355 nm light compared to profiles predicted by others. Intensity is normalized by the maximum droplet intensity, and the horizontal dimension is normalized by the droplet radius.

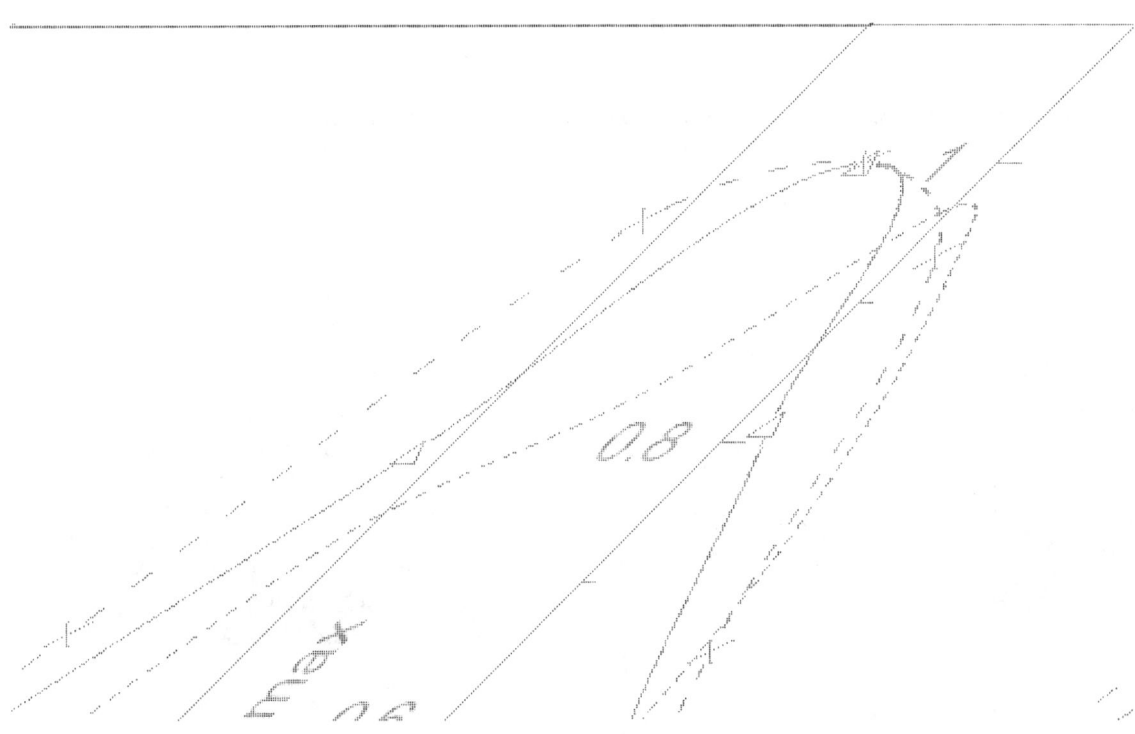

Figure 32. Horizontal intensity profile through the centerline of a 200 μm nominal diameter water drop illuminated with 355 nm light compared to predicted profiles[9]. Intensity is normalized by the maximum droplet intensity, and the horizontal dimension is normalized by the droplet radius.

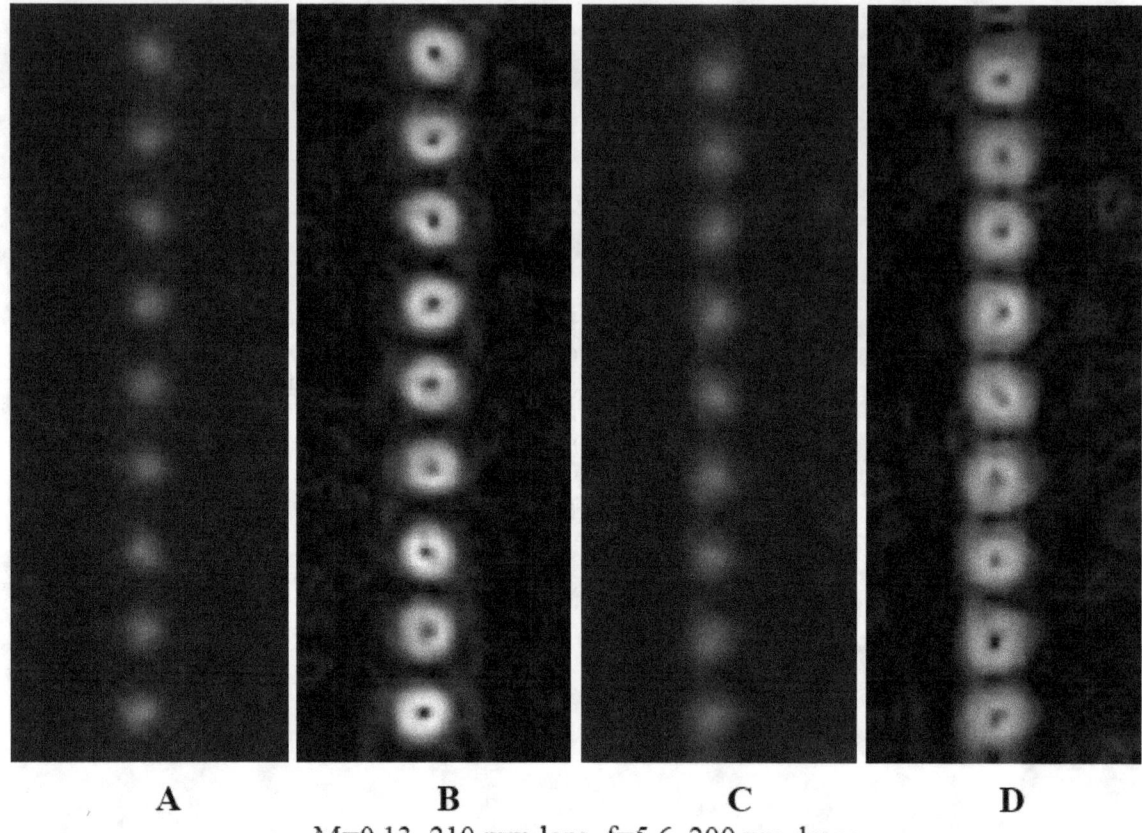

A   **B**   **C**   **D**

M=0.13, 210 mm lens, f=5.6, 200 um drops

Figure 33. Examples of 200 μm nominal diameter droplets. Here, 'A' and 'B' are illuminated with 532 nm light, while 'C' and 'D' are illuminated with 355 nm light. Also, 'A' and 'C' are raw images of droplets. The result of applying the sobel filter to these images is shown in 'B' and 'D'. Note that the images here are limited in quality by the electronic file size limitations of publication.

R65 small droplet and sobel profile blue

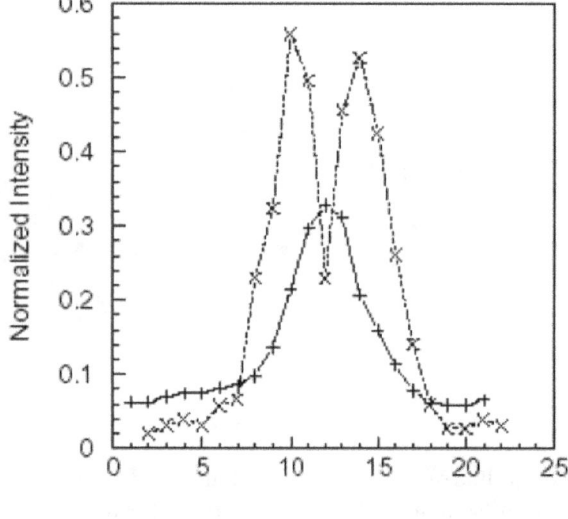

+———+ Inorm,blue
X------X Inorm,blue,sobel filter

Figure 34. Normalized intensity profile of 200 μm nominal diameter water droplet conducted through the image centerline before and after application of the sobel filter. The droplet was illuminated with a 355 nm laser sheet. The intensity is normalized by the maximum scanner intensity, and plotted against distance in pixels.

R65 small droplet and sobel profile yellow

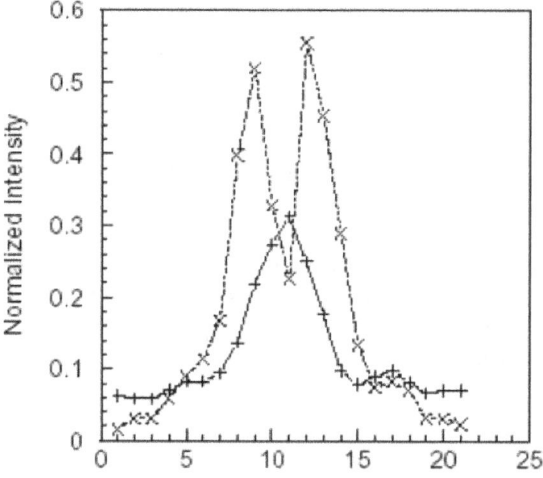

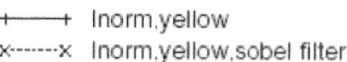
+———+ Inorm,yellow
x------x Inorm,yellow,sobel filter

Figure 35. Normalized intensity profile of 200 μm nominal diameter water droplet conducted through the image centerline before and after application of the sobel filter. The droplet was illuminated with a 532 nm laser sheet. The intensity is normalized by the maximum scanner intensity, and plotted against distance in pixels.

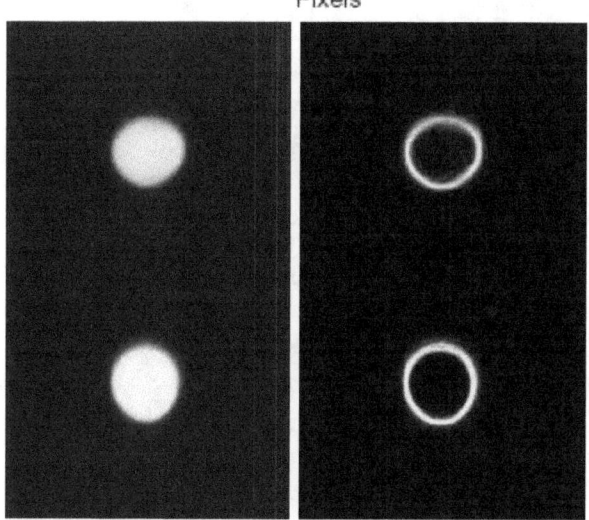

Figure 36. Examples of a 3 mm nominal diameter falling drop. The uppermost droplet images in both panes are the result of illumination with 532 nm light, while the lower droplet images are from 355 nm light. The left pane contains raw images of the drop, while the results of applying the sobel filter to these images are shown in the right pane. Note that the images here are limited in quality by the electronic file size limitations of publication.

R85E05 large droplet and sobel profile blue

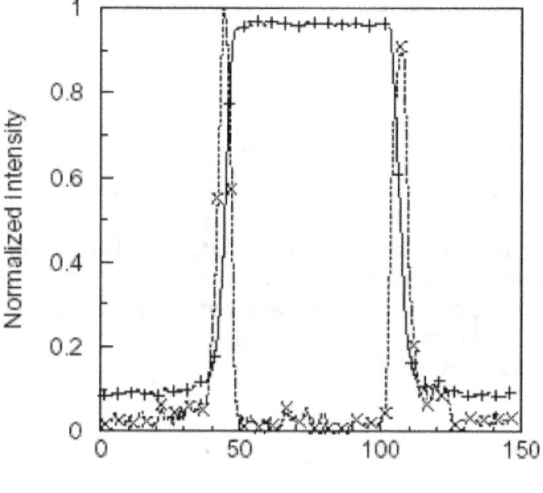

+———+ Inorm,blue
x------x Inorm,blue,sobel filter

One out of every five data points shown for clarity

Figure 37. Horizontal intensity profile of 3 mm nominal diameter water droplet conducted through the image centerline before and after application of the sobel filter. Droplet was illuminated with a 355 nm laser sheet. The intensity is normalized by the maximum scanner intensity, and plotted against distance in pixels.

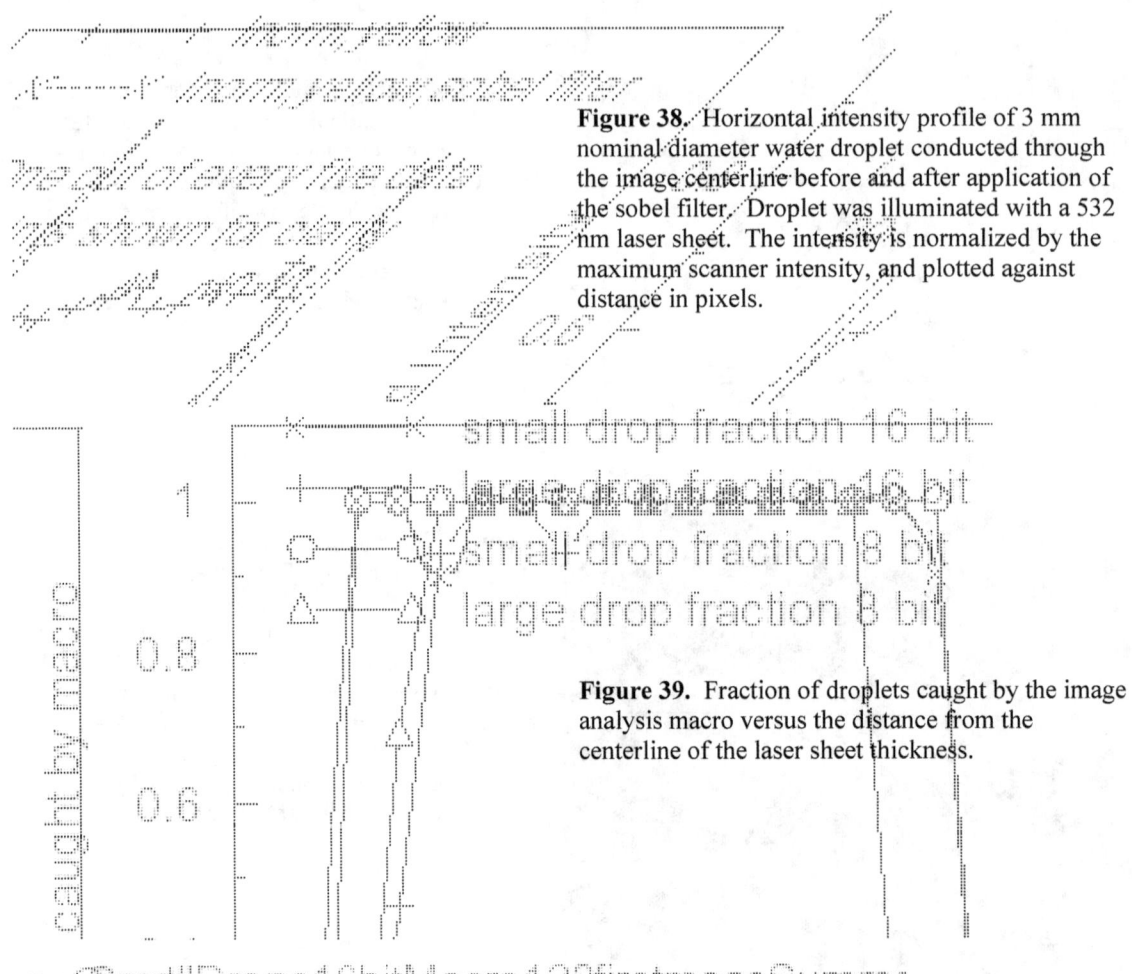

Figure 38. Horizontal intensity profile of 3 mm nominal diameter water droplet conducted through the image centerline before and after application of the sobel filter. Droplet was illuminated with a 532 nm laser sheet. The intensity is normalized by the maximum scanner intensity, and plotted against distance in pixels.

Figure 39. Fraction of droplets caught by the image analysis macro versus the distance from the centerline of the laser sheet thickness.

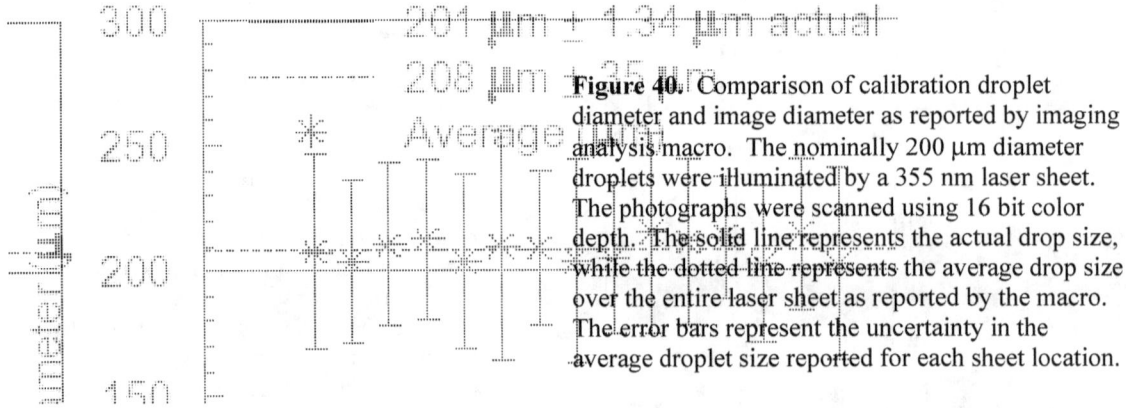

Figure 40. Comparison of calibration droplet diameter and image diameter as reported by imaging analysis macro. The nominally 200 μm diameter droplets were illuminated by a 355 nm laser sheet. The photographs were scanned using 16 bit color depth. The solid line represents the actual drop size, while the dotted line represents the average drop size over the entire laser sheet as reported by the macro. The error bars represent the uncertainty in the average droplet size reported for each sheet location.

Figure 41. Comparison of calibration droplet diameter and image diameter as reported by imaging analysis macro. The nominally 200 μm diameter droplets were illuminated by a 355 nm laser sheet. The photographs were scanned using 8 bit color depth. The solid line represents the actual drop size, while the dotted line represents the average drop size over the entire laser sheet as reported by the macro. The error bars represent the uncertainty in the average droplet size reported for each sheet location.

BigDrops16bitMacro122Summary Graph

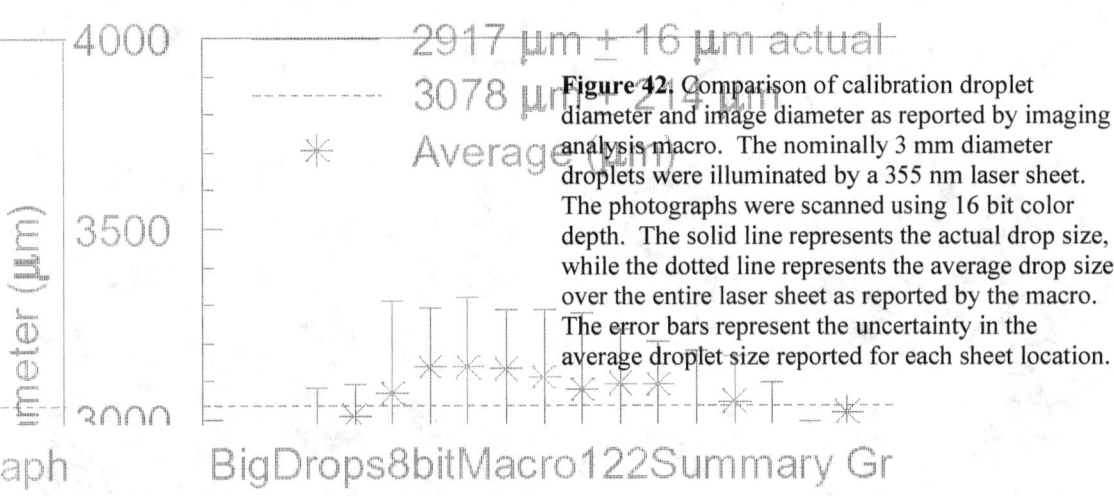

Figure 42. Comparison of calibration droplet diameter and image diameter as reported by imaging analysis macro. The nominally 3 mm diameter droplets were illuminated by a 355 nm laser sheet. The photographs were scanned using 16 bit color depth. The solid line represents the actual drop size, while the dotted line represents the average drop size over the entire laser sheet as reported by the macro. The error bars represent the uncertainty in the average droplet size reported for each sheet location.

BigDrops8bitMacro122Summary Graph

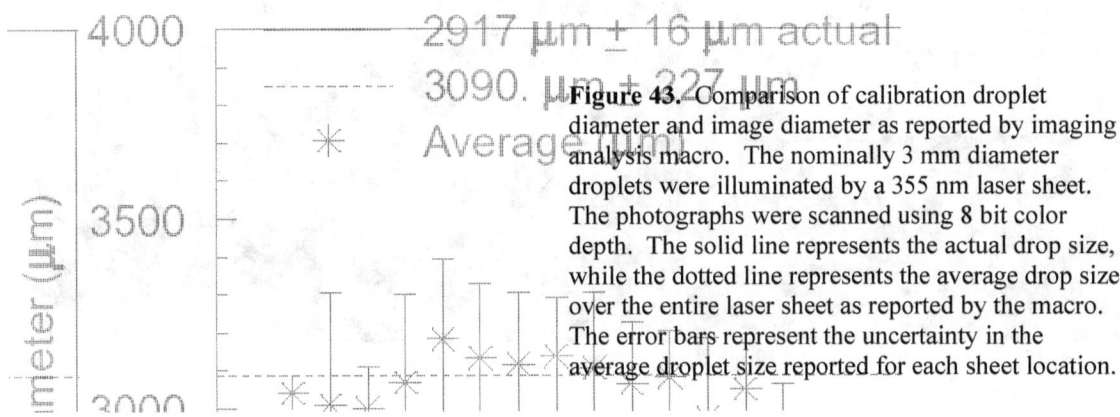

Figure 43. Comparison of calibration droplet diameter and image diameter as reported by imaging analysis macro. The nominally 3 mm diameter droplets were illuminated by a 355 nm laser sheet. The photographs were scanned using 8 bit color depth. The solid line represents the actual drop size, while the dotted line represents the average drop size over the entire laser sheet as reported by the macro. The error bars represent the uncertainty in the average droplet size reported for each sheet location.

Figure 44. Low laser power at 532 nm (10 mJ/shot). High laser power at 355 nm (120 mJ/shot). Imaged area: 84 mm x 118 mm.

Figure 45. High laser power at 532 nm (200 mJ/shot). High laser power at 355 nm (120 mJ/shot). Imaged area: 84 mm x 118 mm.

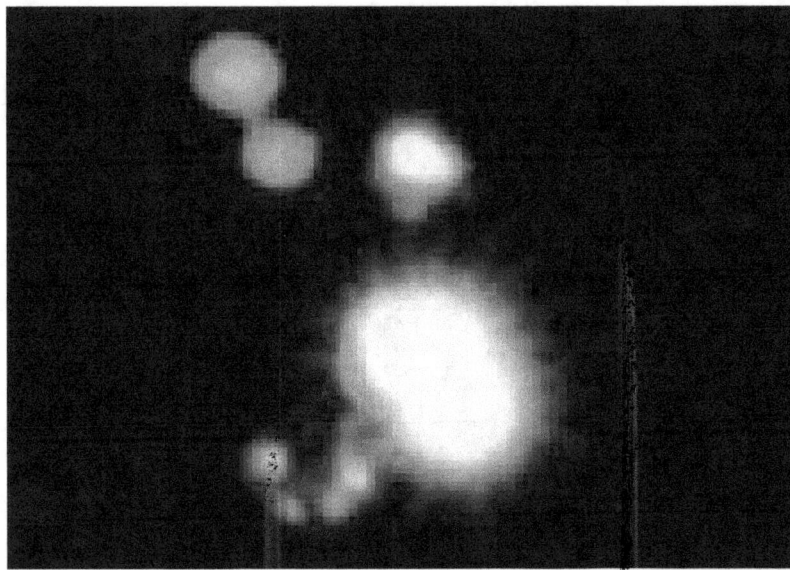

Figure 46. Enlarged view of a small region of figure 45. The upper red drop is caused by UV excitation. Its diameter is 1.47 mm or 21 pixels. The bright yellow drops are caused by visible radiation. A green corona appears due to scattering at the front edge of the drop.

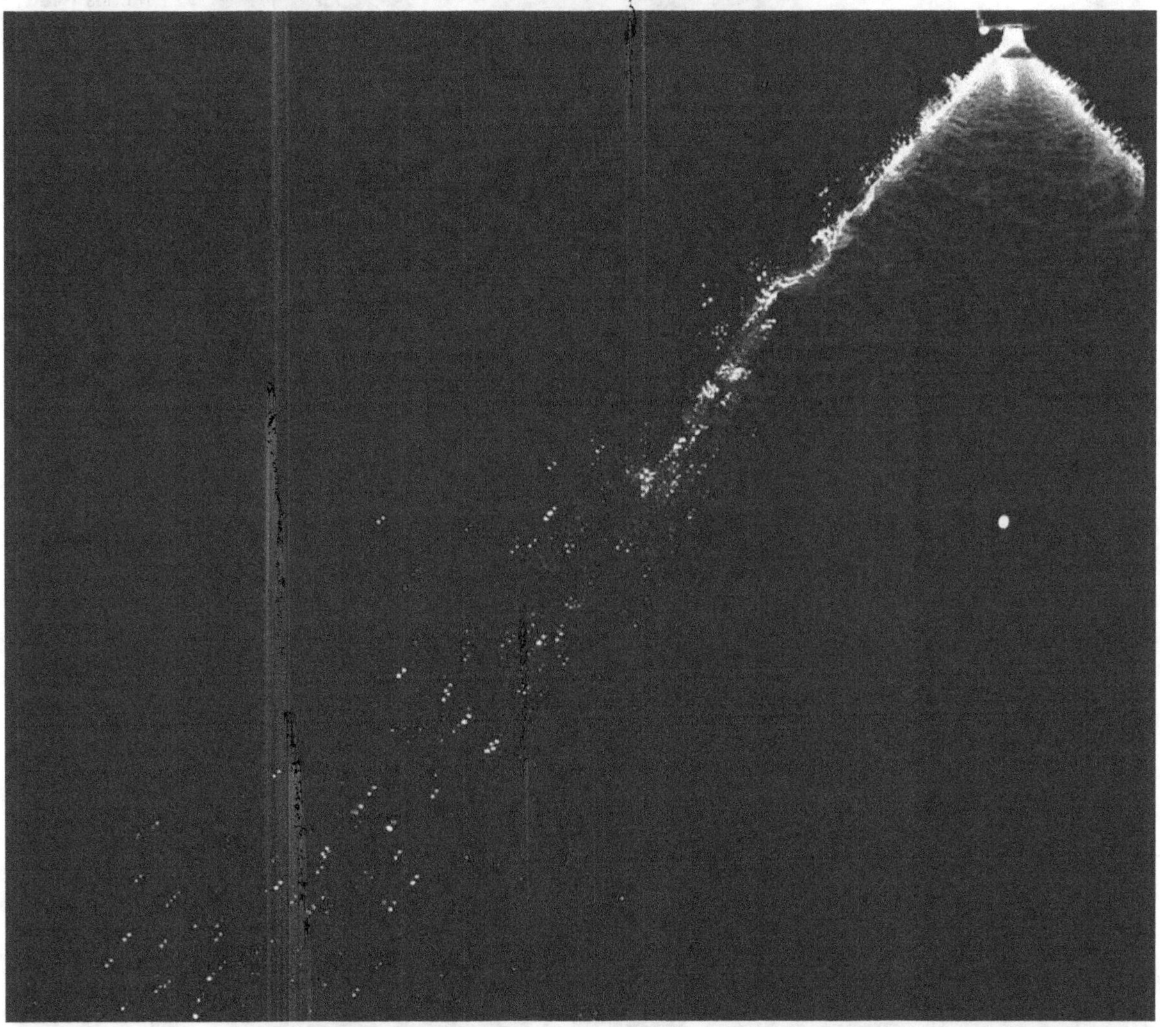

Figure 47. Axis-symmetrical sprinkler with 6 mm nominal diameter orifice and 90 degree strike plate operating at approximately 6 gpm. Image area is approximately 21 inches wide by 17 inches high.

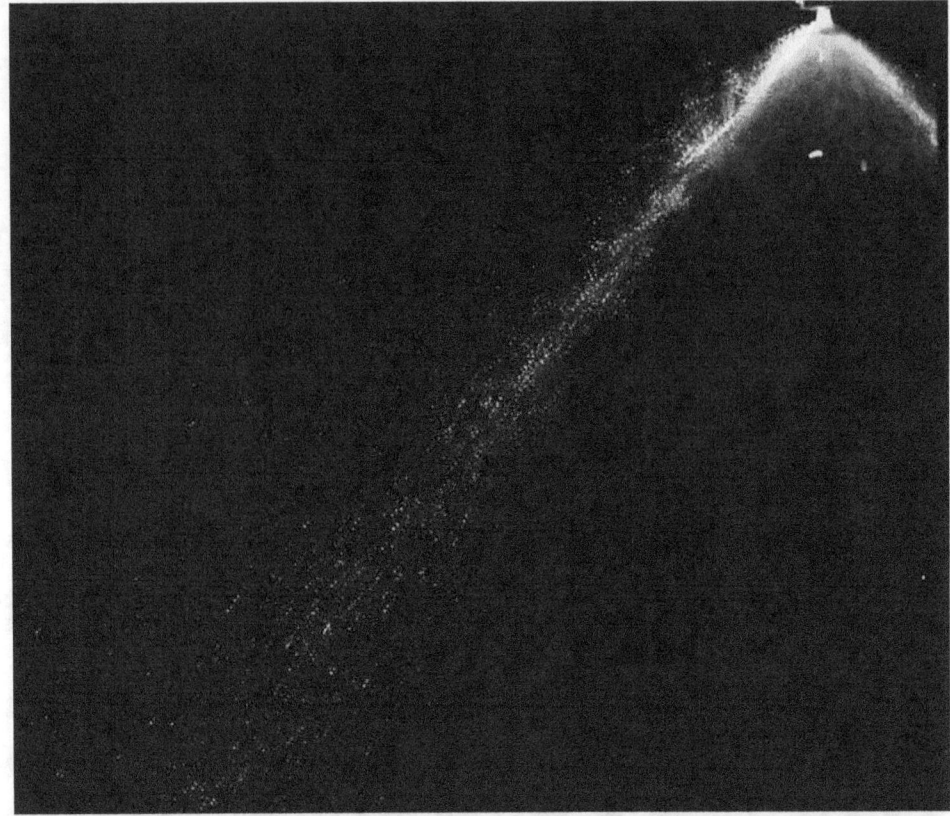

Figure 48. Axis-symmetrical sprinkler with 4 mm nominal diameter orifice and 90 degree strike plate operating at approximately 6 gpm. Image area is approximately 21 inches wide by 17 inches high.

Figure 49. Axis-symmetrical sprinkler with 6 mm nominal diameter orifice and 90 degree strike plate operating at approximately 6 gpm. Image area is approximately 21 inches wide by 17 inches high. The upper right hand corner of the image area is located approximately 8.5 inches vertically down from the orifice, and 5.5 inches horizontally from the symmetric centerline of the orifice. The large orange circle is a light emitting diode used as a reference location.

Figure 50. Axis-symmetrical sprinkler with 4 mm nominal diameter orifice and 90 degree strike plate operating at approximately 6 gpm. Image area is approximately 21 inches wide by 17 inches high. The upper right hand corner of the image area is located approximately 6 inches vertically down from the orifice, and 5 inches horizontally from the symmetric centerline of the orifice. The large orange circle is a light emitting diode used as a reference location

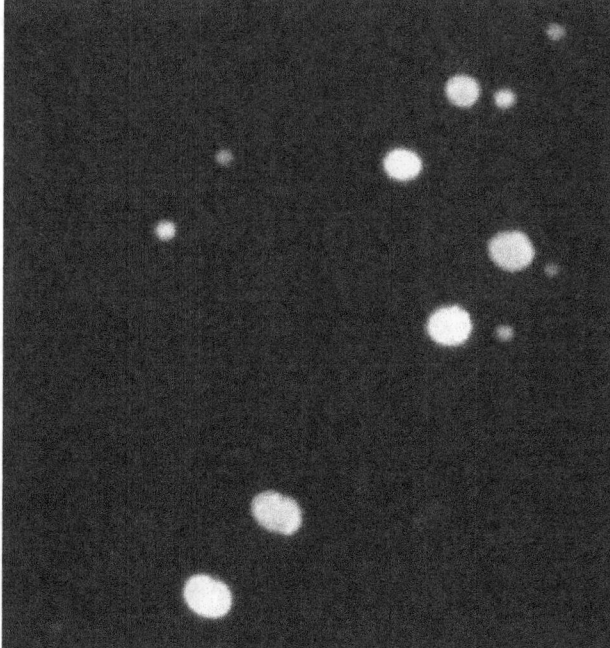

Figure 51. Magnified portion of the measurement area, measuring approximately 20 mm x 20 mm, from the 6 mm nominal orifice diameter sprinkler operating at approximately 6 gpm.

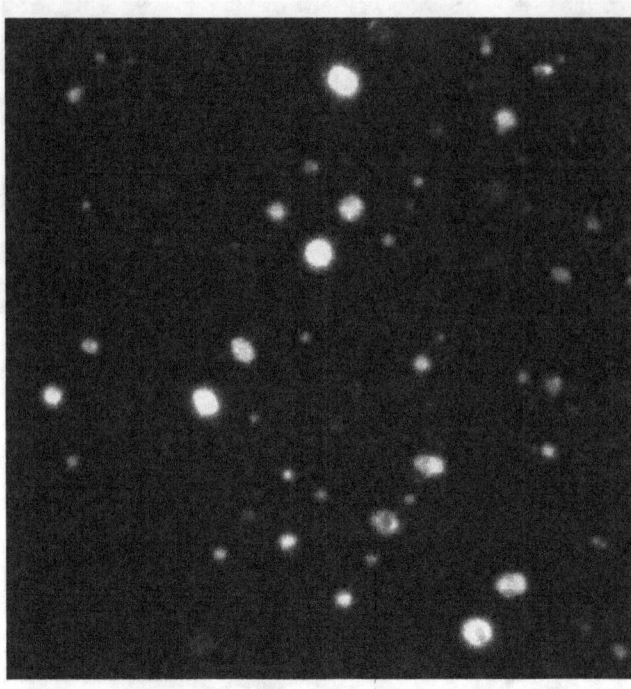

Figure 52. Magnified portion of the measurement area, measuring approximately 20 mm x 20 mm, from the 4 mm nominal orifice diameter sprinkler operating at approximately 6 gpm.

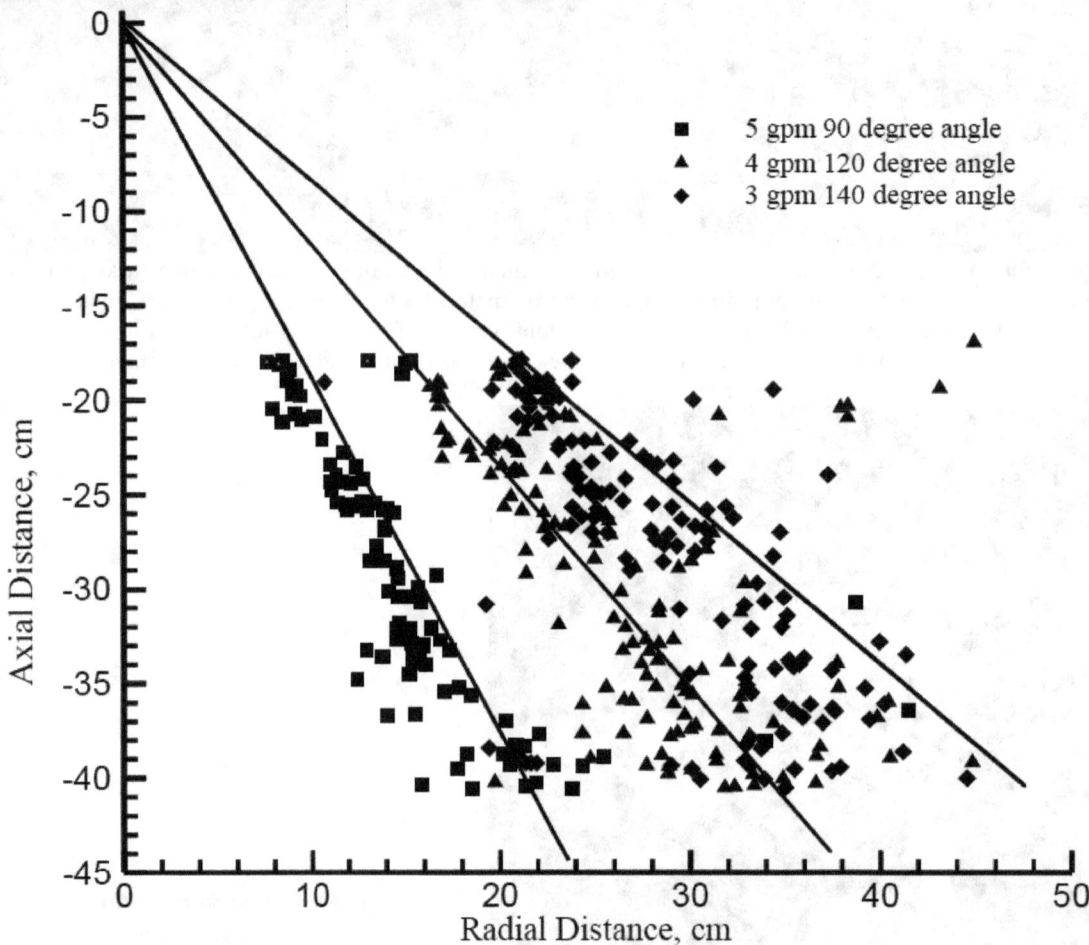

Figure 53. Drop trajectories for a nozzle with an orifice diameter of 8mm. These trajectories are indicated by three lines corresponding to different cone angles of the strike plate.

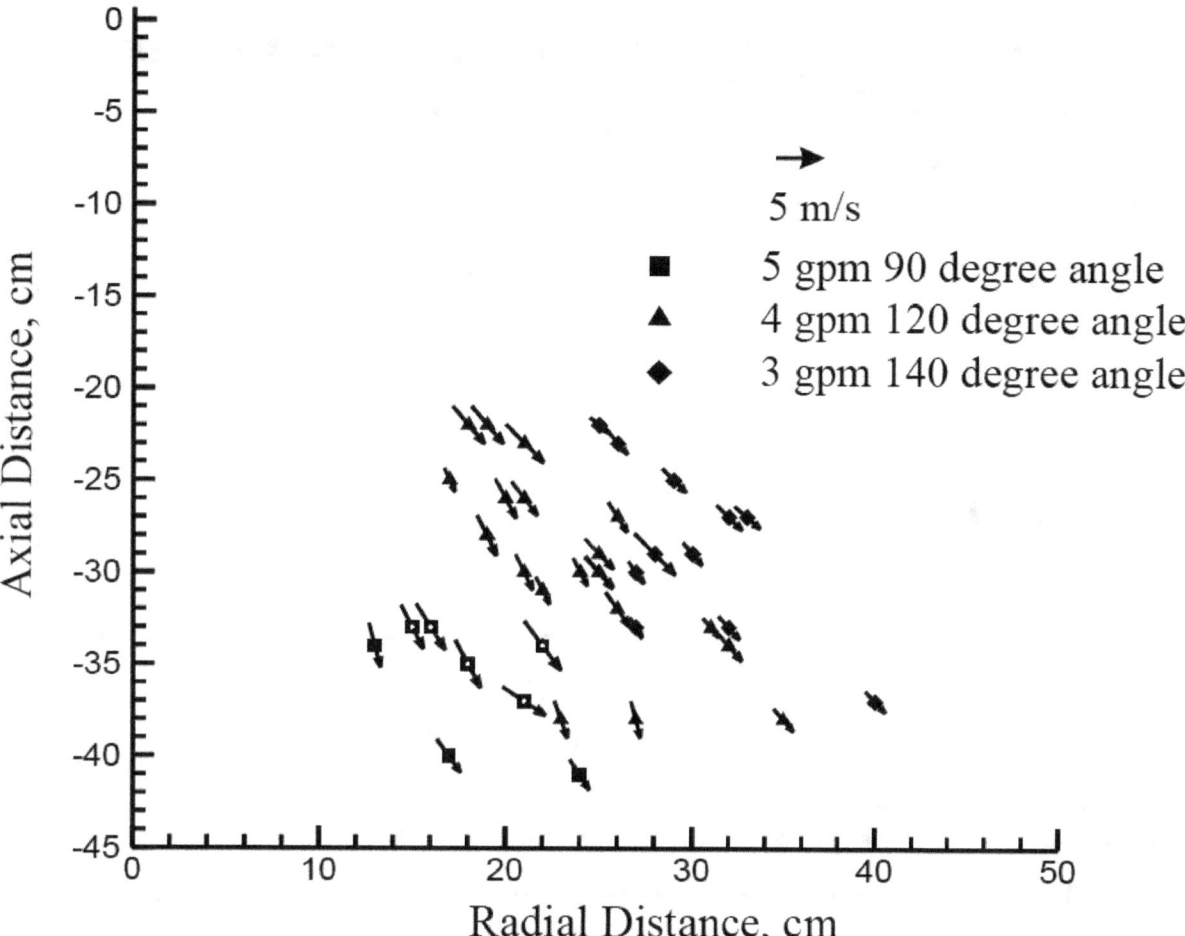

Figure 54. Drop velocities for a nozzle with an orifice diameter of 8 mm for three different strike plate cone angles.

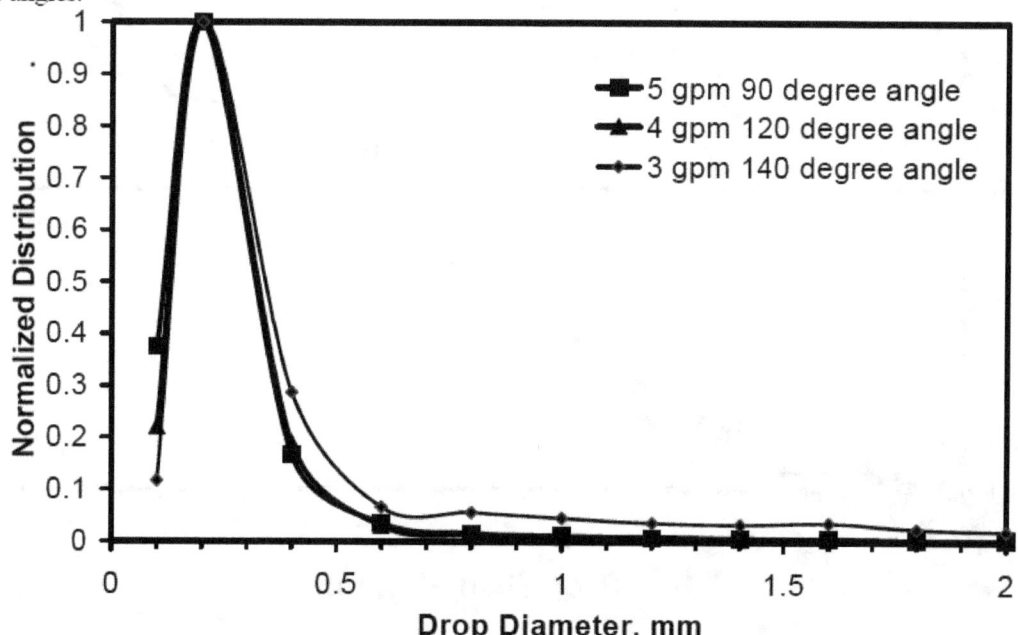

Figure 55. Drop size distribution for a nozzle of 8 mm diameter normalized by the maximum number density.

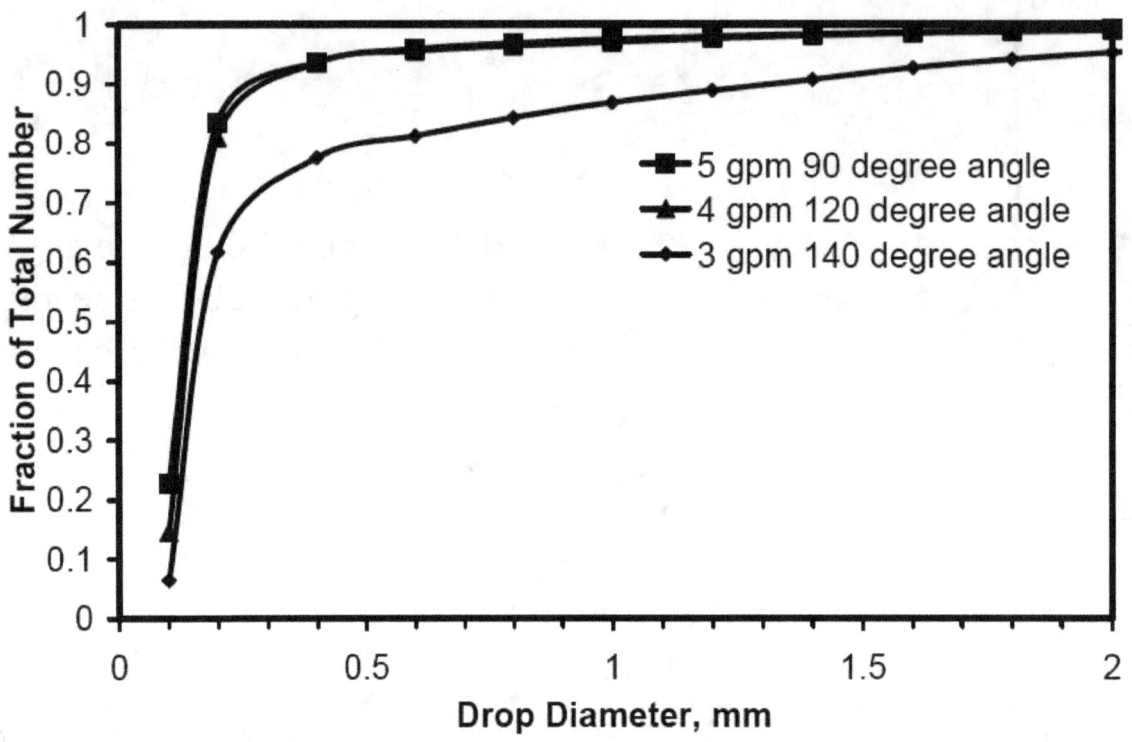

Figure 56. Cumulative number density distribution for the 8 mm diameter nozzle. Thus, comparatively fewer number of drops are larger than 0.5 mm.

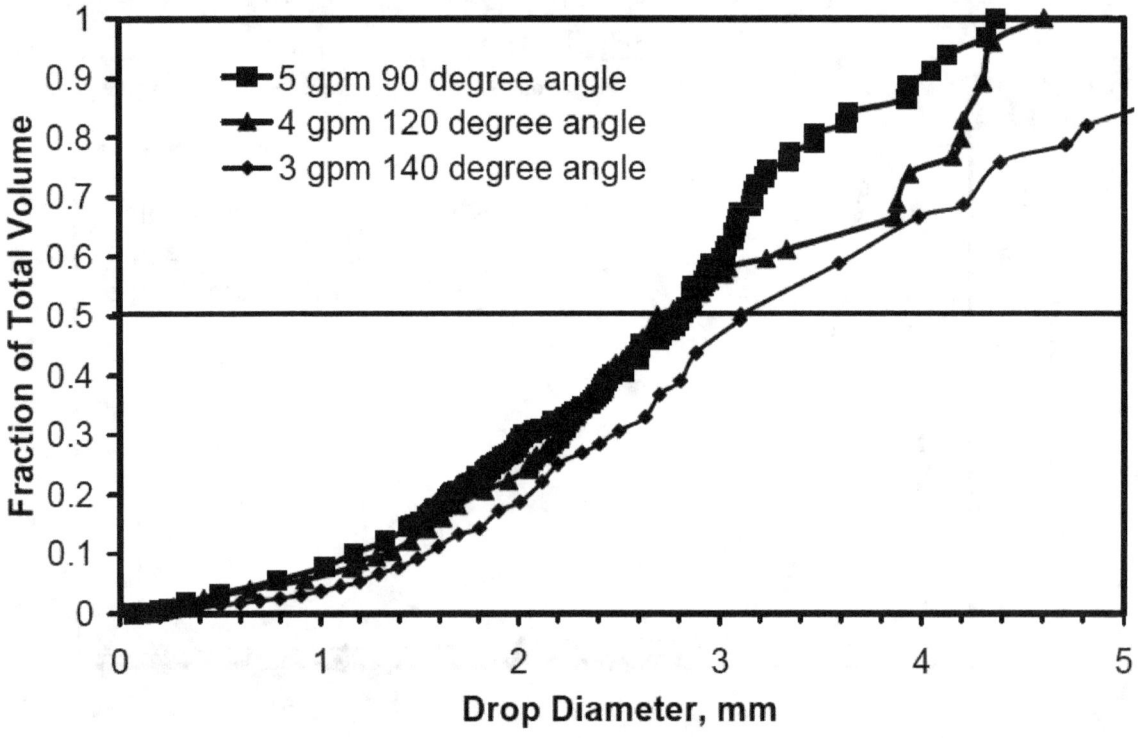

Figure 57. Cumulative volume fraction distribution for the 8 mm diameter nozzle. The drop diameter at 0.5 is 2.8 mm indicating that half the water is carried by drops larger than 2.8 mm.

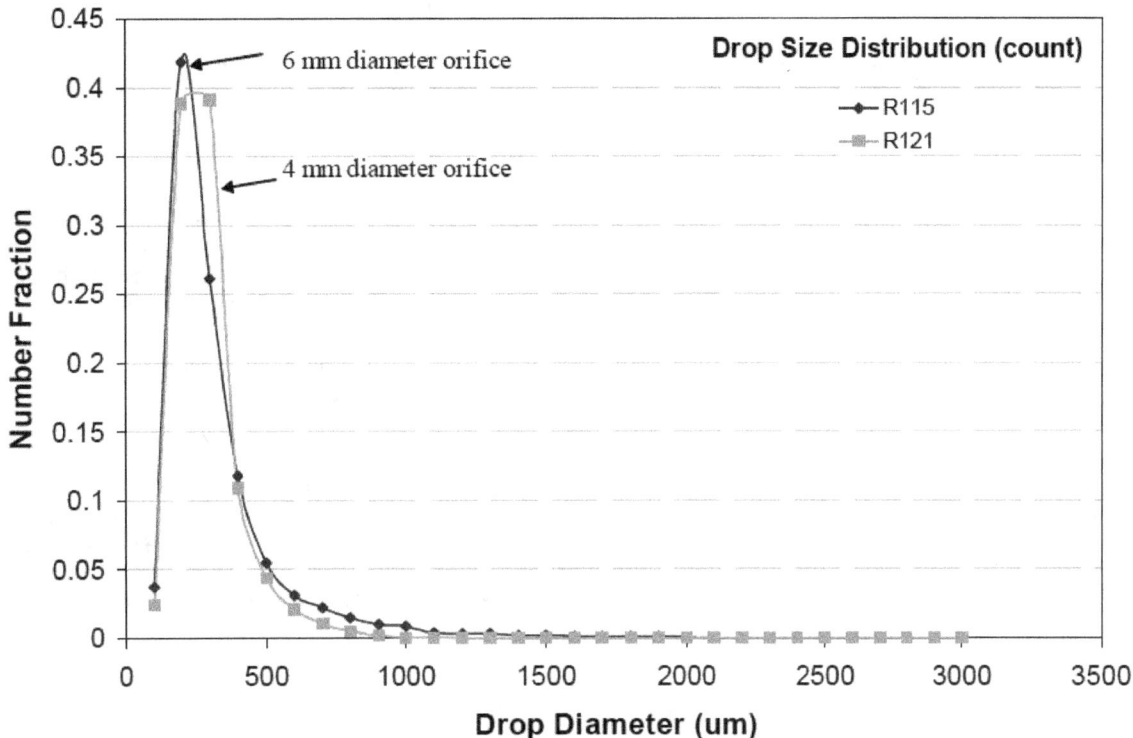

Figure 58. Drop size number fraction distribution is compared for the 6 mm and 4 mm nominal diameter orifice sprinklers operating at 6 gpm.

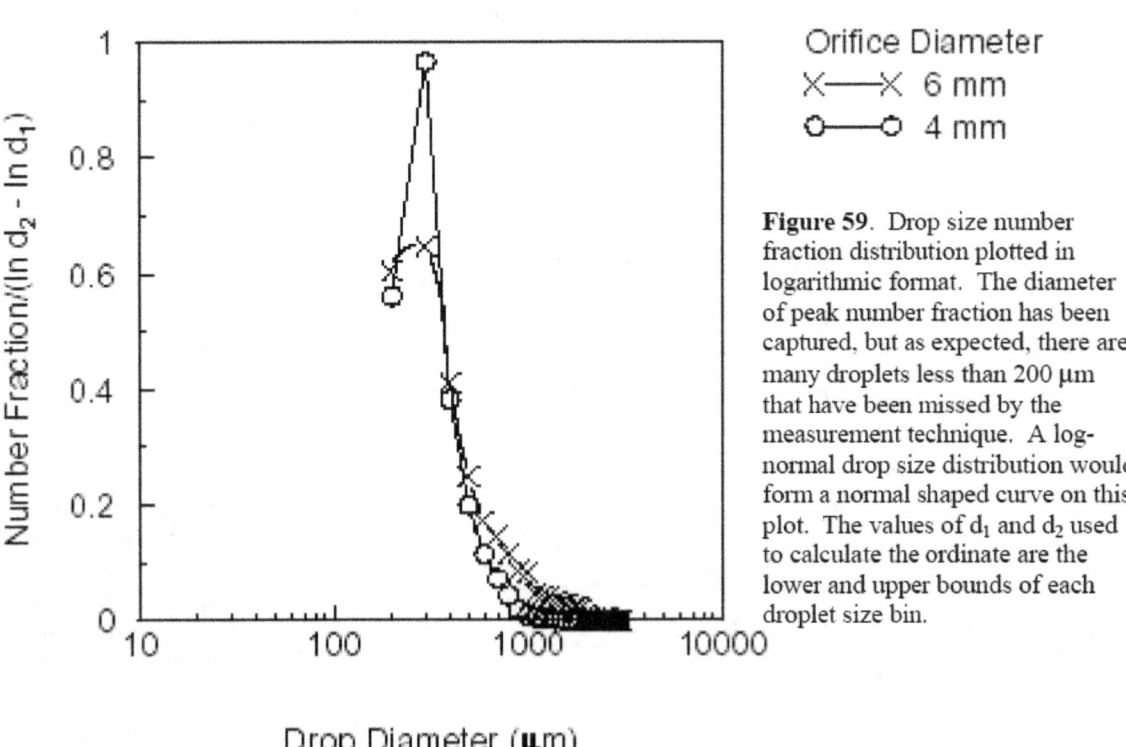

Figure 59. Drop size number fraction distribution plotted in logarithmic format. The diameter of peak number fraction has been captured, but as expected, there are many droplets less than 200 µm that have been missed by the measurement technique. A log-normal drop size distribution would form a normal shaped curve on this plot. The values of d_1 and d_2 used to calculate the ordinate are the lower and upper bounds of each droplet size bin.

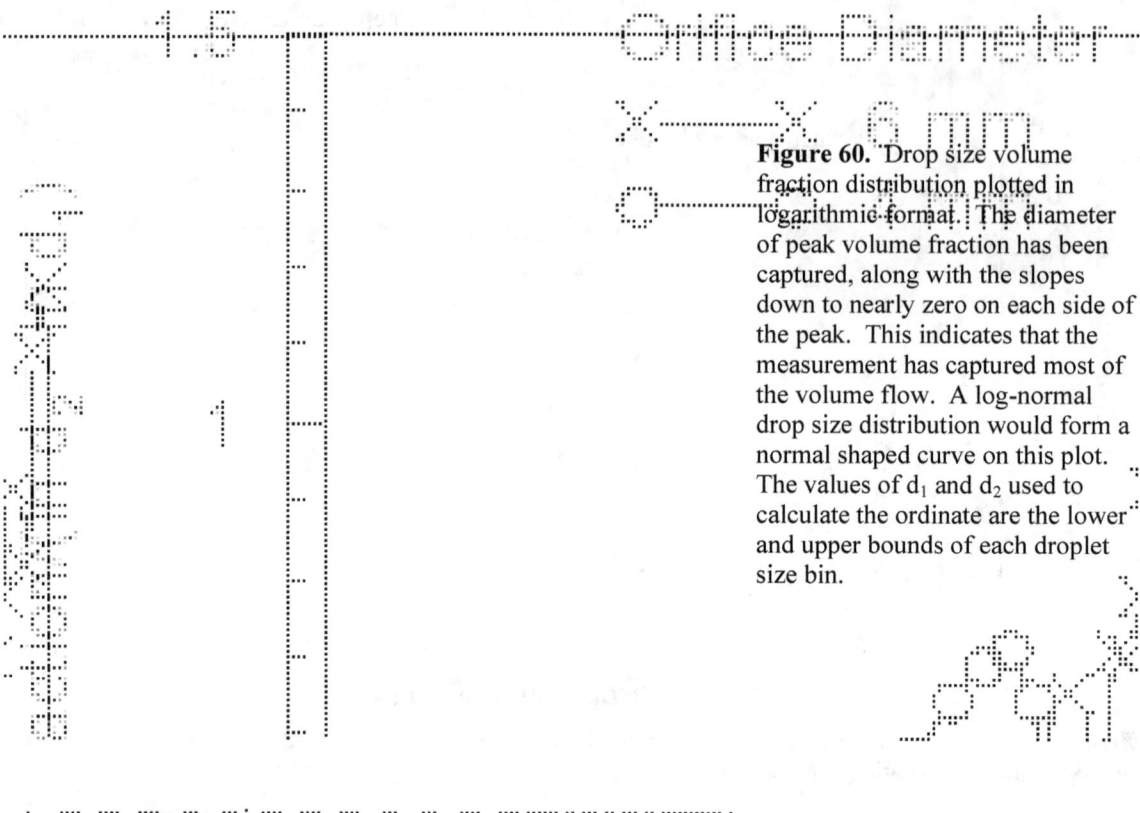

Figure 60. Drop size volume fraction distribution plotted in logarithmic format. The diameter of peak volume fraction has been captured, along with the slopes down to nearly zero on each side of the peak. This indicates that the measurement has captured most of the volume flow. A log-normal drop size distribution would form a normal shaped curve on this plot. The values of d_1 and d_2 used to calculate the ordinate are the lower and upper bounds of each droplet size bin.

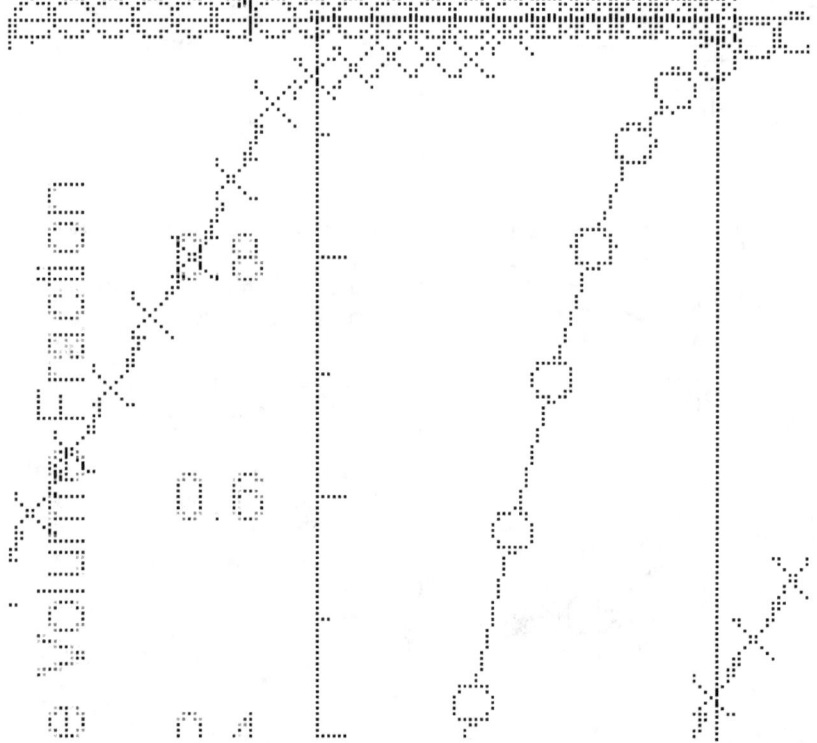

Figure 61. Cumulative volume fraction distribution. The 6 mm nominal orifice sprinkler produces droplets containing a higher fraction of water at larger drop sizes than the 4 mm orifice sprinkler.

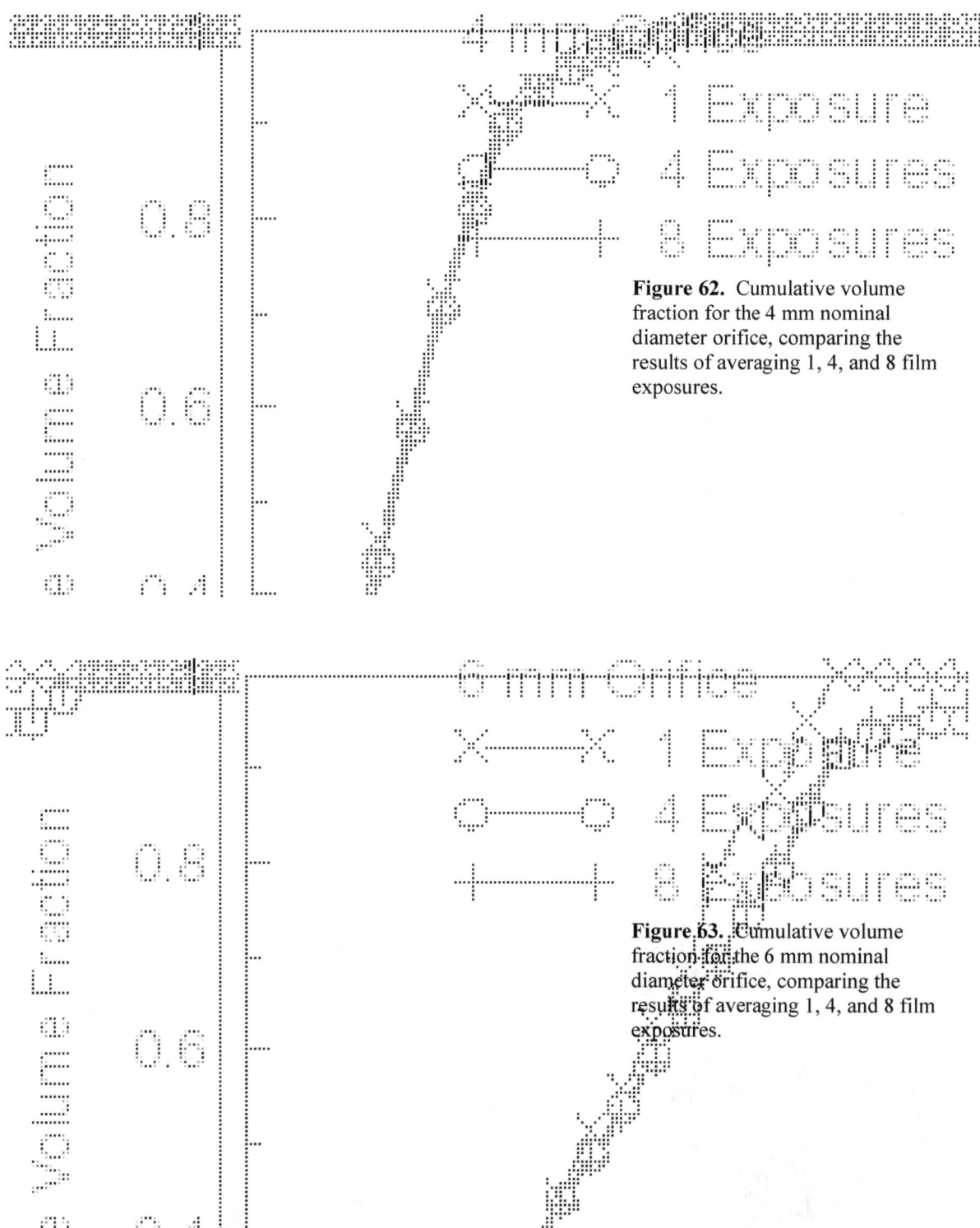

Figure 62. Cumulative volume fraction for the 4 mm nominal diameter orifice, comparing the results of averaging 1, 4, and 8 film exposures.

Figure 63. Cumulative volume fraction for the 6 mm nominal diameter orifice, comparing the results of averaging 1, 4, and 8 film exposures.

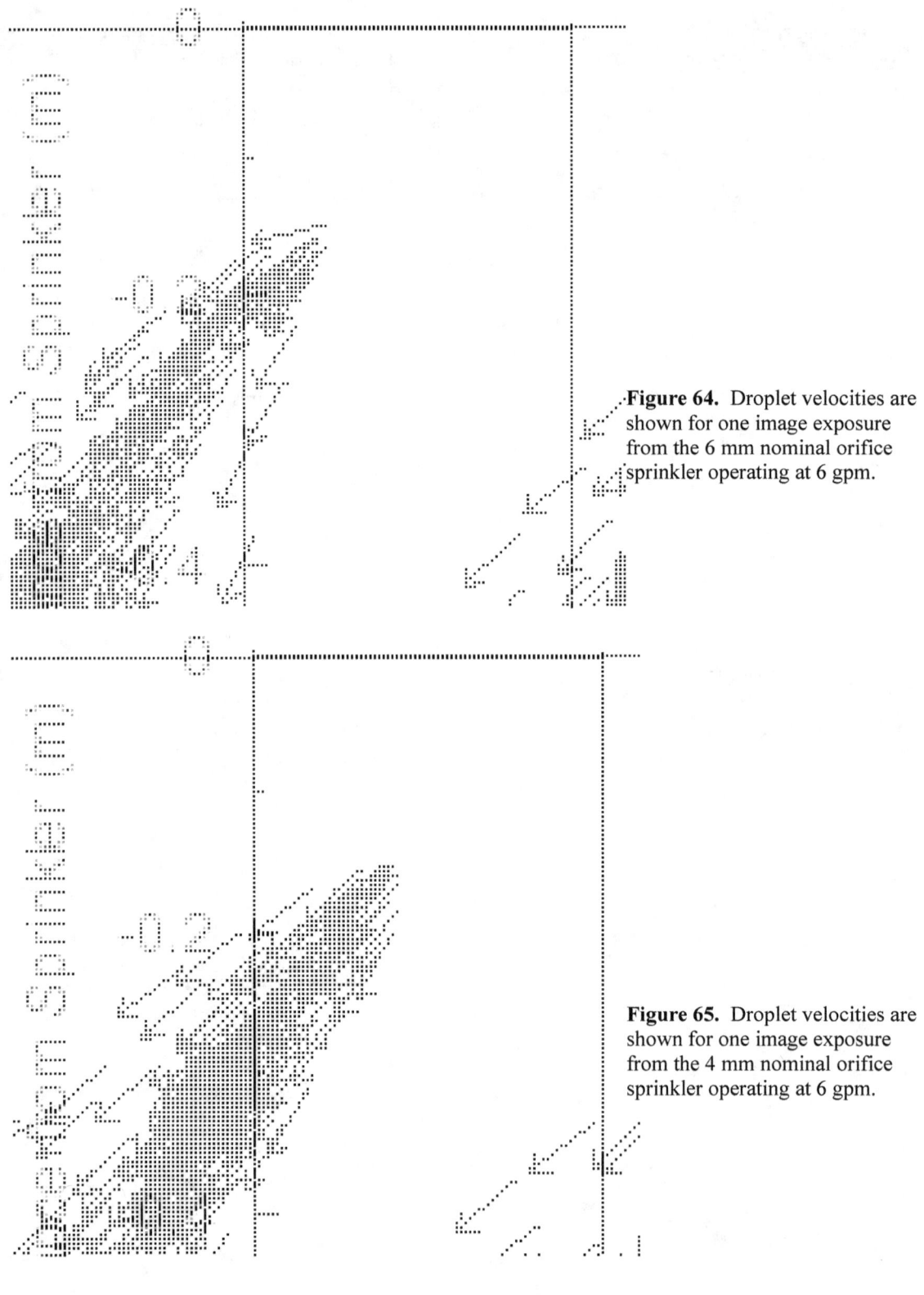

Figure 64. Droplet velocities are shown for one image exposure from the 6 mm nominal orifice sprinkler operating at 6 gpm.

Figure 65. Droplet velocities are shown for one image exposure from the 4 mm nominal orifice sprinkler operating at 6 gpm.

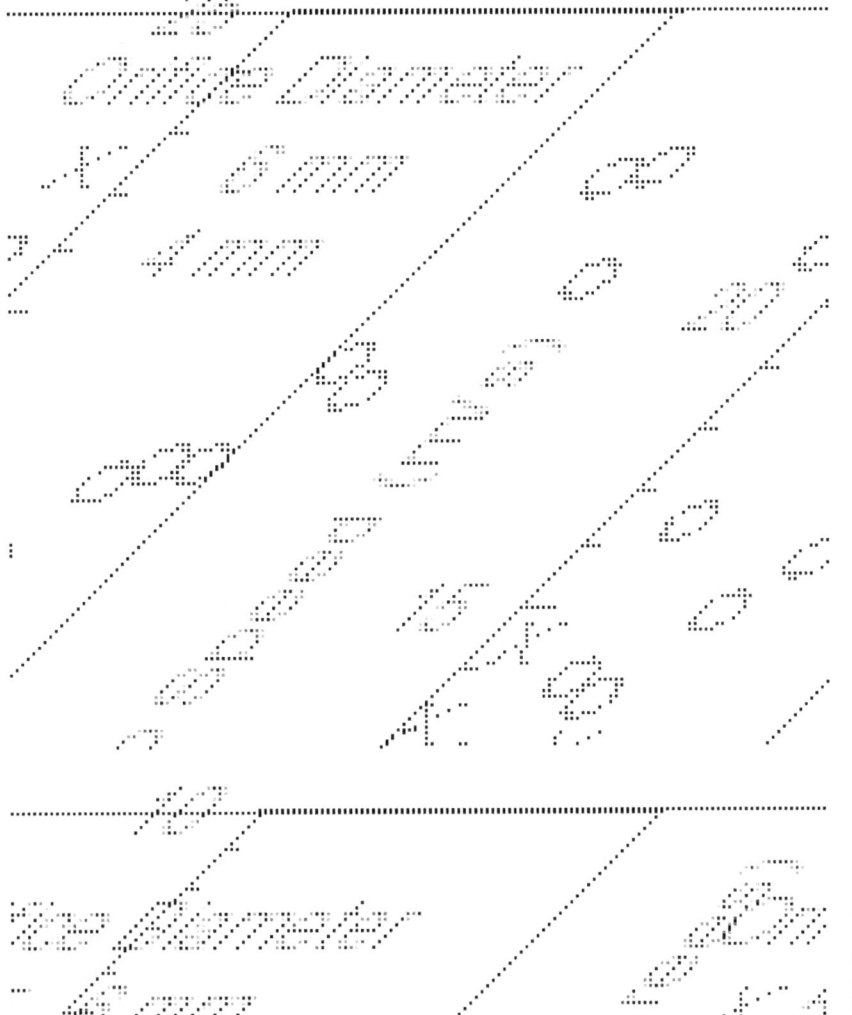

Figure 66. Average droplet speed as a function of drop diameter.

Figure 67. Average droplet trajectory deviation from strike plate angle. The cone angle of the strike plate is 90 degrees, so the strike plate surface is located 45 degrees counter-clockwise from the vertical axis of symmetry (south pole defined as 0 degrees). All of the deviations are negative, indicating that the droplet paths are displaced from the 45 degree reference angle toward the south pole.